e original of this catalog is in the Museum of American Ice
rvests and Woodworking Crafts at Mequon, Wisconsin. The
talog was loaned to The Collector's Door by the museum's
rector, Robert Siegel. This reproduction is intended as a
ference source only.

addition to reprinting tool catalogs, The Collector's Door
blishes a quarterly magazine in the interest of the collec-
ng of antique and unique tools. We also have an extensive
llection of tools available to sell or trade, and offer for
le nearly all in-print books on the subject of old tools
d related crafts and trades.

THE COLLECTOR'S DOOR
854 Maple Avenue
Noblesville, Indiana
46060

ESTABLISHED IN 1834.

WM. T. WOOD & CO.,

MANUFACTURERS OF

FINEST QUALITY

ICE TOOLS

OF EVERY DESCRIPTION.

FACTORY AT ARLINGTON, MASS.

OFFICE AND WAREROOMS:

NO. 49 NO. MARKET STREET,

BOSTON, MASS.

PRESS OF L. BARTA & CO., 54 PEARL STREET, BOSTON.

ANNUAL GREETING.

Our Fifty-fourth Annual Catalogue is hereby presented to our customers and the Ice Tool trade, containing illustrations, descriptions, and prices of the various goods we manufacture, which, we trust, will be a guide to those who contemplate ordering Ice Tools, in making up their memoranda of articles wanted.

Our Stock of Ice Tools is very large and complete. The demand for our goods has grown upon us so rapidly, particularly during the last dozen years, that we have each year been striving to meet it by increasing our stock very largely; and while we have been in some seasons much behind our orders, because they increased in a much greater ratio than our preparations, we have now to offer, as a result of a full year's work with our large force of experienced workmen, and with our improved facilities, the largest stock of Ice Tools ever presented for sale in the history of the ice business.

The Quality of our Tools is the best that the most skilful labor and long experience can produce, starting with the indispensable foundation element—the finest stock. Purchasers of Ice Tools do not place sufficient importance upon the fact that cheaply made Ice Tools are unprofitable to them, and expensive in the end. The use of them serves to give a premium to the manufacturer of cheap tools, who, by selling them at only a small reduction from the price of well-made goods, makes a much larger profit than good tools can command, while the purchaser is also damaging his own interests by obliging his men to use inferior tools; for in no business is a more critical test required of its working implements than in that of harvesting and handling ice.

Our long experience in meeting the demands of the largest and most exacting ice men in the country, and our unwavering disposition to work for the best possible results to our customers, has placed us in a leading position as manufacturers, which is very gratifying to ourselves, and which we are determined to maintain at whatever cost, by the continued manufacture of the

FINEST QUALITY OF ICE TOOLS ONLY.

We use the best cast steel; Norway iron in all parts requiring strong and solid work, and double-refined iron in all other parts. Even the difference between this stock and that which is but a trifle inferior, such as is used by some good makers, is very perceptible to the experienced ice man, and the reputation we enjoy for making fine tools is largely due to our close adherence to the use of this high grade of stock, as well as to the careful manner in which it receives treatment in manufacturing, and to the fact that we steadily engage a score of the most experienced workmen in our special and difficult branch of edge-tool making that are to be found, and who have been engaged with us for years.

Our Prices are as low as strictly first-class work can be afforded, and for what we give for the money, we believe they are the cheapest Ice Tools in the market; for we have unequalled facilities for manufacturing at the lowest possible cost, and we are therefore confident that if tools can be bought cheaper than what we offer them for, they are not equal to ours in quality or serviceability.

Our Office and Warerooms are now in Boston, at 49 NORTH MARKET STREET, where we carry a full stock of all varieties of our goods, and all who are interested in Ice Tools or Elevating Machinery, are invited to call upon us there, where a corps of experienced salesmen will be found in readiness to attend to all inquirers.

Personal Attention in both manufacturing and selling is secured, inasmuch as Mr. Cyrus Wood, our senior member, constantly superintends the work at the factory, while Mr. William E. Wood is to be found at the Boston salesrooms.

Purchasers of Seeds, Agricultural Implements and Machines, Woodenware, &c., will find an immense assortment at 49 North Market Street, where Parker & Wood of which firm our Mr. W. E. Wood is senior member, conduct a flourishing wholesale and retail business.

Telephone Communication between the office in Boston and the factory enables us to transmit immediately any special instructions or specifications for odd or unusual tools, as well as to obtain any desired specific information, so that customers enjoy all the advantages at the store which even the factory itself could afford. ☞ Telephone at Boston office is No. 1467.

Our Elevator Work is represented by sample at our warerooms in Boston, and parties contemplating purchase can here examine the style and strength of that which we sell.

Testimonials of the quality and worth of our manufactures we have never printed. They can be had anywhere, and from almost any one in the ice business. For over fifty years our goods have been distributed over all of the United States and ice-producing or ice-handling foreign countries, and to-day the size of our trade — which greatly exceeds that of any other Ice Tool manufacturers — testifies to the merit of our work.

Permit us to take this occasion to thank our customers for their past favors, and to assure them that we shall be solicitous in future, as heretofore, to see that their interests are well served.

To those who may never have used any of our tools, we give invitation to try them, and prove by their handiness and service that our claims are not vain.   Yours truly,

WM. T. WOOD & CO.

49 NORTH MARKET STREET, BOSTON, MASS.
September 1, 1888.

REPAIRING.

IN SHIPPING TOOLS FOR REPAIRS, PLEASE MARK THEM THUS:

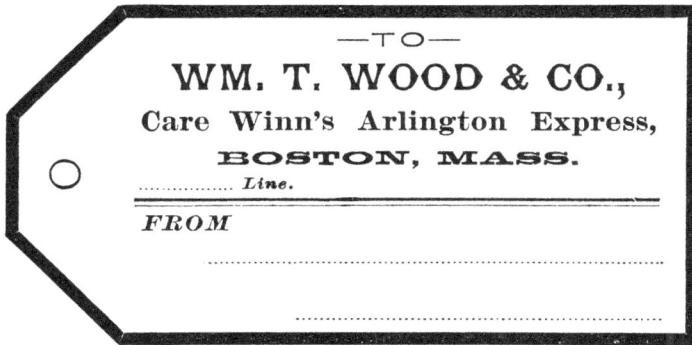

—TO—
WM. T. WOOD & CO.,
Care Winn's Arlington Express,
BOSTON, MASS.
.... *Line.*

FROM

OBSERVE THE FOLLOWING DIRECTIONS:

1st. Send by some through line. The freight will be less, and the goods will come quicker.

2d. Send us the bill of lading.

3d. Fill out the space after " From" on the tag. Several lots often reach us at once.

4th. Always notify us when you ship.

WE REPAIR ALL KINDS OF ICE TOOLS, WHETHER OF OUR OWN MAKE OR OTHERS, PROMPTLY, WELL, AND AT LOWEST PRICES. IT PAYS THE USER TO GET HIS REPAIRING DONE AT OUR FACTORY, INSTEAD OF ALLOWING THE LOCAL BLACKSMITH TO DO IT.

Markers that have become worn too shallow for profitable use can be made as good as new by the insertion of **New Teeth**. The cost is about $25. *Always send the guide, that it may be fitted accurately.*

Plows also sometimes need new teeth, when worn too shallow. Generally, however, plows can be made like new—excepting their depth—by re-forging. This process serves to restore the teeth to their *proper gauge* or width of cut, so that the deeper ones will follow easily,—all being graded. It also re-tempers and re-finishes the teeth and makes the plow good for many years of service, if properly used, before more repairs are needed. Re-forging, including painting, usually costs about $7.00.

Re-Steeling, a job frequently necessary to be done to iron and steel plows, is generally worth about $9.00. Gumming out (filing and painting included), usually $2.50 to $3.00.

Burned Plows or Markers can be made as good as new by working over, and an expense of $11.00 or $12.00 will cause a pile of junk to be transformed into a nicely painted and serviceable ice plow, with new case. Such plows or markers must have been originally of good make and of good stock, in order to be worth much after repairs. We cannot make a first-class article out of a cheaply made tool.

The Standard Gauge for Ice Plows is designated in fortieths of an inch, but is approximately as follows : Marker, 7-16 inch ; 6-inch and 7-inch Plows, 13-32 inch ; 8-and-9 inch Plows, 3-8 inch, full; 10-inch Plows, 11-32 inch, full; 12-inch, 15-32 inch, scant; 14-inch, 5-16 inch.

The great proportion of all Ice Plows in the country are made to this gauge when new,—they tend to wear narrower,—and it is desirable to keep one's plow sufficiently near the proper gauge that the purchase of a new plow will not disarrange the order of their use.

Chisels and Bars which have become worn up, so as to be of but little use, can be made as good as new by re-steeling or welding on new ends.

Ice Hooks can be overhauled, and by drawing out, or by supplying new pullers, shovers, or handles when needed, can be made into good hooks for a season's use, or more, at less than half the price of new ones.

Prices for New Teeth in plows will be sent on application. Prices for clearing teeth on page 48.

Prompt attention to all repairs. Satisfaction guaranteed.

WM. T. WOOD & CO.

ICE TOOLS.

In no business, perhaps, is it more necessary to have its working tools perfect in every respect than in the ice-cutting business. The cutting season is short, the weather is cold, the men are hurried, and an unnecessary delay on account of unmanageable or imperfect tools is both aggravating and expensive. It is, therefore, of the greatest importance to the ice dealer that he should possess the *best* tools that can be bought. Quantities of cases can be cited where users of cheap tools, bought because they were cheap, have lost many times the cost of good ones by delays, breakages, and poor service.

Our aim in manufacturing is, **1st,** to make the best in the market; **2d,** to sell at the lowest price that the best can be sold.

We have the following advantages in favor of carrying out this aim :—

We are the oldest Ice Tool manufacturers in the United States, and consequently have the longest experience.

We have the largest and most extended trade in Ice Tools of any manufacturers, consequently are subject to a greater scope of suggestions.

We manufacture Ice Tools only, giving our entire time and energies to the perfecting of them alone.

We have unequalled facilities for manufacturing, by means of which we can execute work rapidly and in a superior manner.

We employ the most skilful workmen, many of whom have been with us for long periods of years.

Our goods are manufactured under our direct supervision, and some of the most particular branches are executed by ourselves personally, thus enabling us the more confidently to guarantee satisfaction.

We have the best of opportunities for making practical tests of our work.

We buy no inferior materials, using only that quality which by experience we know will make the best article.

We purchase in large quantities, at the lowest market rates for cash, and import our own steel.

We carry an extensive stock of manufactured goods, and can fill large orders promptly.

In view of the above, we feel confident of being able to give complete satisfaction to all, and we hope to place in the hands of every dealer a sample at least of our goods.

WM. T. WOOD & CO.,

BOSTON, MASS

FILING PLOWS AND MARKERS.

REMARK.—Every ice dealer should have a good wooden straight-edge, five and a half feet long, by which to keep his plow filed level.

DIRECTIONS.

In filing a plow it will be well to observe the following method:

FIRST. File the point of each tooth (as at A) sharp and square, doing all of the filing on the bottom of the tooth (that is, the upper side when the plow is turned upside down), and never filing on the front of the tooth in the space between it and the next tooth (as over E), except to turn the feather edge down toward the ice.

SECOND. Place the long straight-edge on the points of the teeth in order to see if they range level, and if not, file the points which are too high until all the points touch the straight-edge.

THIRD. File each heel (if necessary) so that the point next behind the heel to be filed will have a strong ¼ inch feed (as indicated at C). Any straight piece of wood or thin iron ten inches long, will answer for a gauge by cutting a notch on the end by which to measure the feed.

FOURTH. Next see that a good clearance be filed directly back of the edge of each tooth (as shown by the space between D and the gauge C), so that the extreme end of the point only will touch the gauge.

FIFTH. Keep the hind heel (F) filed so that it will measure about the same distance to the beam of the plow as the point (A) of the same tooth.

On Clearing-Tooth Plows, keep the clearing-tooth **very blunt** so as not to cut, and let it be kept filed down so that when the long straight-edge is placed upon the points of the cutting-teeth, there will be about ⅛ inch space between the point of the clearing-tooth and the straight-edge.

To increase the feed of plows or markers, file a little off of each heel (including the hind heel); to decrease the feed, file a little off of each point.

A well-made Ice Plow filed according to above directions **cannot fail** to cut.

Sometimes, however, a plow may appear to be in perfect order, having a proper feed and sharp points, and still will not cut well. This is usually because the feather edges are not turned down. By simply rubbing the points on the upper side with the back of a jack-knife blade, so as to cause the fine edges to project downward toward the ice, such a plow will be made to cut its full amount. This device is often sufficient to cause a partially dull plow to cut well.

A FILING GAUGE, *having a notch on the end by which to file the feed, will be furnished free to any customer who may send for one.*

WM. T. WOOD & CO.

CONTRACTORS' SUPPLIES.

A FULL LINE OF THE CELEBRATED

COOK, RYMES & CO. PATTERNS

CONSTANTLY IN STOCK.

Since we purchased the Ice Tool and Paving tool business of Messrs. COOK, RYMES & Co., in April, 1886, we have been manufacturing a full variety of their well-known brand of goods. *The best stock only* is used in them, and their quality is far superior to the ordinary articles generally found in the market.

THEIR SUPERIORITY IS UNIVERSALLY RECOGNIZED.

We illustrate a few of the principal tools below.

C. R. & CO. PICK. Regular weight, 9 lbs.

LIGHT PAVING HAMMER. Best Cast Steel.

HEAVY PAVING HAMMER. Best Cast Steel.

WOODEN PAVING RAMMER. Wt. 35 to 50 lbs.

CONTRACTORS' MAUL.

STEEL-FACED IRON PAVING RAMMER. Weight, 45 to 50 lbs.

SPIKING HAMMER.

EXTRA QUALITY STEEL CROWBAR. 6 ft. Long, Round or Square Point.

PRICE LIST.

Wood Rammers, each	$4.00	Heavy Paving Hammers, each	$2.00
Steel-Faced Rammers, each	8.00	Railroad Adzes, each	2.50
C. R. & Co. best Picks, not hdld., doz.	21.00	Spiking Hammers, 6 to 7 lbs., each,	2.00
,, ,, ,, ,, handled ,,	24.00	Crowbars, extra steel, 6 ft., each	2.00
Solid Eye Best Picks, not hdld., doz.	15.00	Tamping Bars, each	2.00
,, ,, ,, ,, handled ,,	18.00	Contractors' Mauls, handled, each,	2.50
Light Paving Hammers, each	2.00	Beetles, handled, each	2.00

SPECIAL TOOLS MADE TO PATTERN AT REASONABLE PRICES.

WM. T. WOOD & CO.,

49 North Market Street, - - - - - Boston, Mass.

ICE WAGON.

It is of great importance to the ice dealer that his wagons be of the most serviceable manufacture that can be obtained.

We are now prepared to furnish our customers with such as will be found first-class in every respect, and at as low prices as is consistent with the highest grade of work.

Our wagons are made from thoroughly seasoned Eastern ash and oak, the best brands of American and Norway iron, and we use only the celebrated

CONCORD SPRINGS AND AXLES.

in their construction.

The workmen employed have had years of experience on the best class of work, and we can confidently assure the Ice Trade that the wagons we offer

CANNOT BE EXCELLED.

The above cut represents the Boston Pattern with Platform Springs.

We also make Three-Spring and Dead-Axle Wagons, which come much less in price.

Special Patterns, will be made to order, and we will furnish, *free of charge,* drawings and designs of wagons in accordance with the views of purchasers who are decided as to the style best adapted to their respective localities.

Supply Wagons, Ice Sleds, and Sleighs also manufactured.

PRICES.

ONE HORSE WAGONS.				TWO HORSE WAGONS.			
Boston Pattern, capacity,	1,500 lbs.,	$200.00		Boston Pattern, capacity,	4,000 lbs.,	$275 00
,,	,,	,,	2,000 ,,	225.00	,,	,,	,,	5,000 ,,	300.00
,,	,,	,,	3,000 ,,	250.00	,,	,, extra heavy ,,	5,000 ,,	325.00
Brakes, $10.00 extra.				,,	,,	,,	6,000 ,,	350.00

ICE ELEVATORS.

We are now prepared to offer our customers the best style of Ice Elevating Machinery manufactured. The use of steam power for raising ice is now very generally adopted, even by those whose storage capacity does not exceed 1,000 or 2,000 tons, and the advantage of a "friction-clutch," by which the chain may be started or stopped instantly, independently of the engine, is fully recognized and conceded. It is almost universally adopted when new machinery is ordered, and in many instances it has been connected with the old style of elevator.

We have two styles of friction-clutches: the "Hudson River," so called, and the James' patent. The "Hudson River" friction has been in use for a great many years, and is the most powerful friction known.

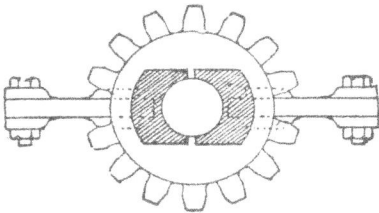

The James' Patent Friction Gear is of recent invention, and while its power is no more than the "Hudson River," it possesses the merits of greater simplicity and compactness, with less weight and cost. It was first introduced to the trade in 1882, and thirty-five were put to the severest test in the following close winter of 1882-83, while many more, put up in the succeeding two winters, have been pleasing scores of customers. Their success has been greater even than was anticipated. We are warranted in the belief that it will be *the* friction-clutch of the future, because of its *extreme simplicity, compactness, durability, and power.*

Our Elevators are made in the most thorough manner, and will be found first-class in every respect.

Our prices we believe to be lower for the quality and weight of work given than those of any others.

We invite a careful comparison of these Elevators in points of weight and strength, size of pulleys, gears, shafts, etc., with any other Elevators in the market.

Please send for our complete ELEVATOR CIRCULAR, describing Undershot and Overshot Elevators, construction of same, details of Friction Clutches, kind of chain used, speed of chain, extras required, etc.

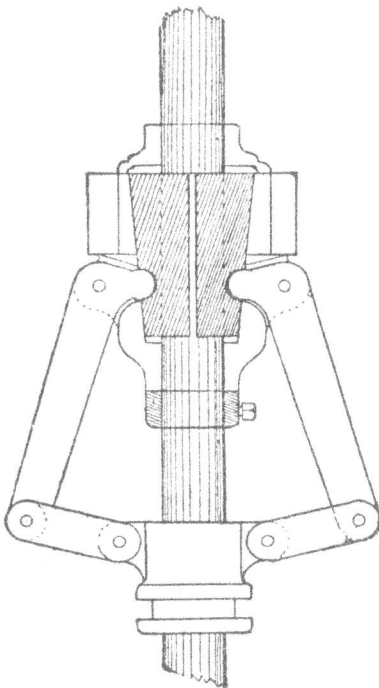

WM. T. WOOD & CO.,

49 North Market St.,

BOSTON.

HENRY JAMES' PATENT FRICTION CLUTCH.

PRICE LIST OF ICE ELEVATORS.

1888-1889.

No. 1. DOUBLE OVER-SHOT ELEVATOR, with "Hudson River"
Friction. Extra heavy. Weight about 4,250 pounds . $330.

For an incline 60 inches in the clear.
To raise *four* 22 X 22 inch cakes on each hold-bar. This consists of the
following pieces :
Friction Shaft, 3⅞ inches, with friction arrangement, pinion 16 X 4½ inches,
and pulley 48 X 12 inches.
Chain Shaft, 3⅞ inches, with chain and ratchet wheels, and gear 56 X 4½ in.
Tram Shaft, 2¾ inches, with tram wheels 20 inches.
Apron Shaft, 2½ inches, with apron wheels 20 inches.
Lever, fulcrum, collars, pillow-blocks, and slide-blocks for tightening the chain.

No. 2. DOUBLE UNDER-SHOT ELEVATOR, with "Hudson River"
Friction. Extra heavy. Weight about 4,100 pounds . $300.

Specifications the same as for No. 1, except that the tram shaft and wheels are
not needed, and the apron wheels are 36 inches in diameter.

No. 3. DOUBLE OVER-SHOT ELEVATOR, with "Hudson River"
Friction. Weight about 3,950 pounds . . . $300.

For an incline 52 inches in the clear.
To raise two 22 X 32 inch cakes on each hold-bar.
Specifications the same as No. 1, except that the friction and chain shaft are
3⅜ inches, tram shaft 2½ inches, and apron shaft 2¼ inches.

No. 4. DOUBLE UNDER-SHOT ELEVATOR, with "Hudson River"
Friction. Weight about 3,800 pounds . . . $275.

Specifications the same as No. 3, except that the tram shaft and wheels are not
needed, and the apron wheels are 36 inches in diameter.

No. 5. DOUBLE OVER-SHOT ELEVATOR, with the James Patent
"Friction Gear." Weight about 3,230 pounds . . $265.

For an incline 60 inches in the clear.
To raise two 22 X 22 or 22 X 32 inch cakes on each hold-bar. This in-
cludes :
Friction Shaft, 3⅜ inches, with friction gear 11 X 4½ inches, and pulley
48 X 12 inches.
Chain Shaft, 3⅜ inches, with spur gear 56 X 4½ inches, chain and ratchet
wheels.
Tram Shaft, 2½ inches, with tram wheels 20 inches.
Apron Shaft, 2¼ inches, with apron wheels 20 inches.
Lever, fulcrum, collars, pillow-blocks, and slide-blocks **for tightening chain.**

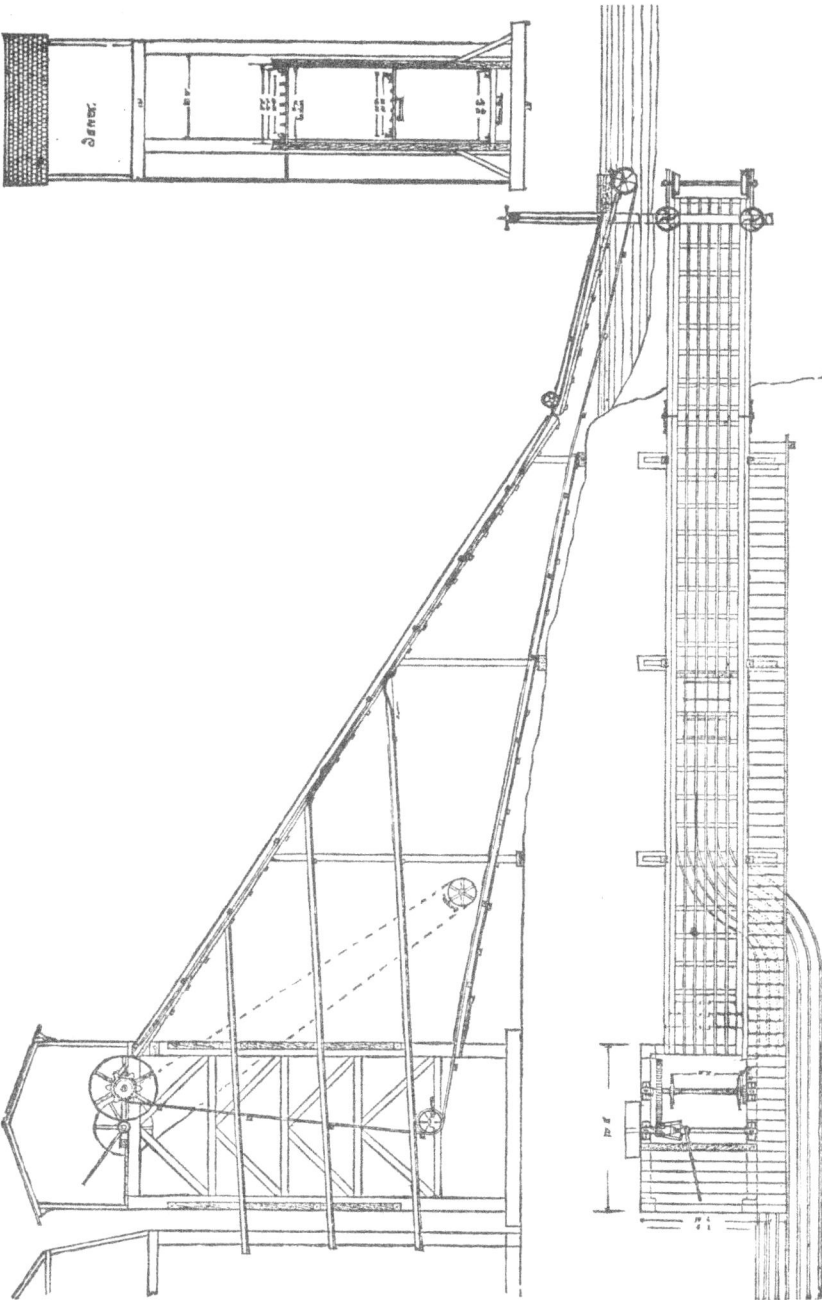

OVER-SHOT ELEVATOR WITH JAMES' PATENT FRICTION GEAR.

No. 6. DOUBLE UNDER-SHOT ELEVATOR, with the James "Friction Gear." Weight about 3,100 pounds . . . $240.

Specifications the same as No. 5, except that the tram shaft and wheels are not used, and the apron wheels are 36 inches in diameter.

No. 7. SINGLE OVER-SHOT ELEVATOR, with the James Patent "Friction Gear." Weight about 2,400 pounds . . $220.

For an incline 44 inches in the clear.
To raise one 22 × 22 or 22 × 32 inch cake on each hold-bar.
Specifications the same as No. 5, except that the friction shaft is 2 15-16 inches, tram shaft 2¼ inches, friction-gear 9 × 4 inches, spur-gear 42 × 4 inches, and pulley 36 × 10 inches.

No. 8. SINGLE UNDER-SHOT ELEVATOR, with the James "Friction Gear." Weight about 2,375 pounds . . . $200.

Specifications same as No. 7, except that the tram shaft and wheels are not needed, and the apron wheels are 36 inches in diameter.

No. 9. DOUBLE OVER-SHOT ELEVATOR, without Friction Clutch. Weight about 2,900 pounds . . . . . $210.

For an incline 60 inches in the clear.
To raise two 22 × 22 or 22 × 32 inch cakes on each hold-bar. Comprising:
Counter Shaft 2 15-16 inches, with pinion 11 × 4½ inches, and pulley 48 × 12 inches.
Chain Shaft 3⅜ inches, with gear 56 × 4½ inches, chain and ratchet wheels.
Tram Shaft 2½ inches, with tram wheels 20 inches.
Apron Shaft 2¼ inches, with apron wheels 20 inches.
Collars, pillow-blocks, and slide blocks for tightening the chain.

No. 10. DOUBLE UNDER-SHOT ELEVATOR, without Friction Clutch. Weight about 2,750 pounds . . . . . $190.

Specifications the same as No. 9, except that the tram shaft and wheels are not used, and the apron wheels are 36 inches in diameter.

No. 11. SINGLE OVER-SHOT ELEVATOR, without Friction Clutch. Weight about 2,400 pounds . . . . . $165.

For an incline 44 inches in the clear.
To raise one 22 × 22 or 22 × 32 inch cake on each hold-bar.
Specifications the same as No. 9, except that the counter shaft is 2 7-16 inches, tram shaft 2¼ inches, pinion 9 × 4 inches, gear 42 × 4 inches, and pulley 36 × 10 inches.

No. 12. SINGLE UNDER-SHOT ELEVATOR, without Friction Clutch. Weight about 2,250 pounds . . . . . $150.

Specifications same as No. 11, except that the tram shaft and wheels are not needed, and the apron wheels are 36 inches in diameter.

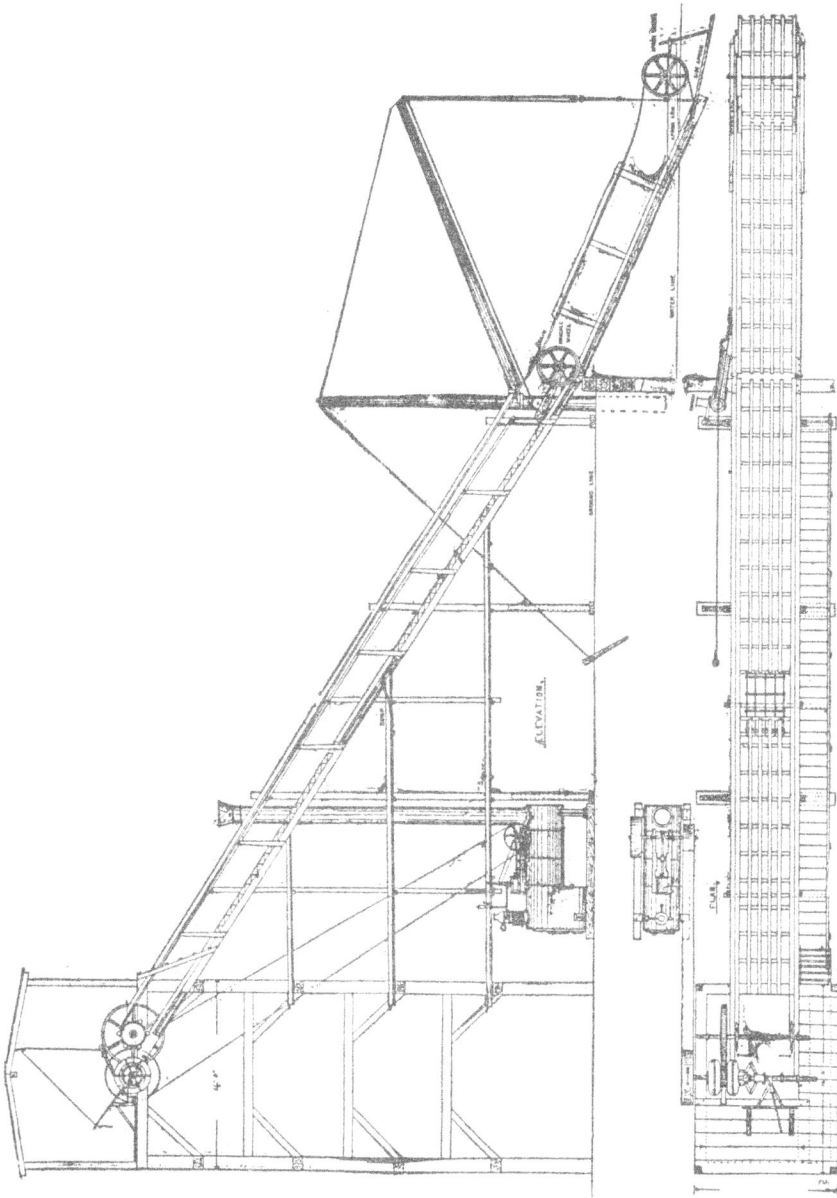

UNDER-SHOT ELEVATOR, WITH "HUDSON RIVER" FRICTION-GEAR.

No. 13. HORSE-POWER OVER-SHOT ELEVATOR, without Friction Clutch . . . . $185.

For an incline 44 inches in the clear.

To raise one 22 × 22 or 22 × 32 inch cake on each hold-bar. It includes :

Chain Shaft 2⅞ inches with three chain wheels, collars and boxes.

Tram Shaft 2¼ inches, with tram wheels 20 inches, collars and slide-blocks.

Apron Shaft 2¼ inches, with apron wheels 20 inches, collars and boxes.

Driving Shaft 3 × 14 feet 6 in., 1 chain wheel, bevel pinion 12 in., and boxes.

Sweep Power, iron frame, shaft 2⅞ in., bevel wheel 36 in., and sweep-block.

(Motion communicated to chain shaft by an ice-chain belt.)

No. 14. HORSE-POWER OVER-SHOT ELEVATOR . . . $90.

Specification the same as No. 13, except the driving-shaft and sweep-power.

No. 15. HORSE-POWER UNDER-SHOT ELEVATOR, without Friction Clutch . . . . $175.

Specifications the same as No. 13, except that the tram shaft and wheels are not used, and the apron wheels are 36 inches in diameter.

No. 16. HORSE-POWER UNDER-SHOT ELEVATOR . . . $80.

Specifications the same as No. 15, except the driving-shaft and sweep-power.

EXTRAS.

Our Ice Chain this year, instead of being made of iron, or common Bessemer steel, as formerly, is now manufactured of the celebrated "Clapp & Griffith" steel, adopted by the U. S., Government in the construction of naval vessels, making a chain that *will not break* under the greatest strain possible in elevating ice.

1. "Hudson" Chain, ⅜ in. steel, ... Per lb. 7¾ cts., per foot, 27 cts.
2. " " 5-16 " " . . . . " " 8 " " " 23¼ "
3. " " ¼ " " . . . . " " 8¼ " " " 19 "
4. "Eastern" " ⅜ " " . . . . " " 7¾ " " " 35 "

For cost of "Hudson" Chain with one pocket and two bolts for each six feet, add 5 cts. per foot extra.

Hold Bars, 3 x 4 in. . . . . Each, 20 to 30 cts.

(Chain delivered in New York or Boston.)

Pockets for hold bars (one to every 6 feet of chain) . . . . each .20
Chain Bolts, for attaching pockets to chain (two to each pocket) . . . . " .05
Iron Dumps or Rockers, for 60 inch or 52 inch incline. . . . . " 10.00
" " " 44 " 52 " " . . . . " 8.00
Lifting Screws, with wheel-nut for raising and lowering apron . . . . per pair 10.00
Apron Hinges for over or under-shot . . . . " 4.00
Apron Straps " " " . . . . " 3.00
Outer Apron Straps and Guides for under-shot only . . . . " 3.50
" Hinges " " " " . . . . " 3.00
Chain Forks, with bolts, for under-shot only . . . . " 4.00
The above forgings, except chain, pockets, and bolts, for over-shot . . . . . . per set 25.00
" " " " " " " under-shot . . . . . . " 35.00
Angle or Knuckle Wheels, 10 inch, and wrist for same . . . . per pair 10.00
" " " 36 " with 2¼ inch shaft, and boxes for same . . . " 30.00
Hinges for the latter 5.00

Send for Complete Illustrated Elevator Circular.

CATALOGUE OF ICE TOOLS.

If you fail to receive a catalogue any season, please send in your name anew.

We publish an ILLUSTRATED ARTICLE on ICE HARVESTING which we shall be pleased to mail free to any applicant.

The cuts we use throughout this catalogue were engraved especially for us from photographs of the tools. While exact copies of many of them appear in other parties' catalogues and advertisements, it is not by purchase or loan from us that they were obtained.

ICE PLOWS.

THE invention of the Ice Plow and its first manufacture took place probably about the year 1829.

The main principle which this invention introduced—that of drawing a succession of teeth over the surface of the ice—being a decided change from the methods previously practised with the saw and the axe, immediately attracted the attention of interested inventive minds; and during the fifteen or twenty years following, experiments were freely made, and much time and money were spent by able mechanics in endeavoring to perfect the machine.

Although many of these experiments were, naturally enough, unsuccessful or impracticable, genuine improvements were rapidly made upon the original Ice Plow.

ORIGINAL PLOW.

This affair was made with wooden beams, and with iron teeth, steeled, and widened or upset at the point only. Next, iron beams of equal width were substituted for the wooden ones. Then followed various alterations, such as flanging the teeth the entire face so as to carry out the chips; making one beam wider than the other, to give room for the chips on one side; cutting out chip spaces in the narrow beam, thus allowing the chips to shoot out freely from between the teeth and permitting the plow to run to its *full depth ;* cutting out the bottoms of the teeth, leaving only the heel and point to be filed ; making the best plows of *steel* teeth, instead of iron and steel ; improving the curve of the teeth ; increasing the degree of finish, etc., etc. ; all of which, although not manifest in the plows of all manufacturers, are developed in the modern Ice Plow as made by us.

Although improvements in the Ice Plow have not been of very frequent occurrence of late years (the Patent Clearing-Tooth invented in 1872 being the only exception of any note), it is more on account of the perfection of the present machine than from any lack of enterprise.

In this connection, notice may well be taken of some of the fruitless experiments above referred to. Among those may be mentioned double markers ; combination plane and marker ; teeth shaped like a lumber marker, slanted back so as to *draw* through the ice ; plows with sixteen teeth ; gouge-pointed teeth ; teeth with grooved faces, called the " lateral cut," and many others.

Although some of these experiments were attended with apparent success, they were all, after a little practical use, abandoned for various reasons, being as a general thing difficult to keep in order, and very unsatisfactory when not in order.

As most of these impracticable machines are now in existence, and as some who participated in these machines are now living, from whom we have obtained the minutest particulars, we have no hesitation in declaring that *our present style of Ice Plow, with their square-pointed teeth, are superior, for practical service, year after year, to any ever invented.*

Previous to 1867, the larger portion of Ice Plows in use below nine inches in depth were the kind commonly known as " Iron Plows "; that is, the teeth were cut out of iron plate and the points were steeled. Upon the addition of expensive machinery to our works at that time, however, we determined to abandon the manufacture of these plows and to adopt the use of the Best Cast Steel only for the teeth, thus making them stiffer, smoother, and more durable.

Steel Plows, however, in order to have a fine cutting edge capable of doing good service, must be made of the *best stock* only. A large number of Ice Plows are for sale all over the country, purporting to be made of best cast steel, which have neither name, reputation, nor merit. Only such Ice Plows as bear the stamp of some first-class manufacturer, who is *known* to use the *best cast steel only*, can be recommended as being profitable to the user.

We have continued this practice of making steel plows only up to the present time ; and combining as they do all of the essential improvements that the experience of over fifty years can suggest, including the valuable Patent Clearing-Tooth for deep plows, we believe we have good reason to pronounce them *the very best* that can be found in the Ice-Tool market.

THE PATENT CLEARING TOOTH
FOR ICE-PLOWS.

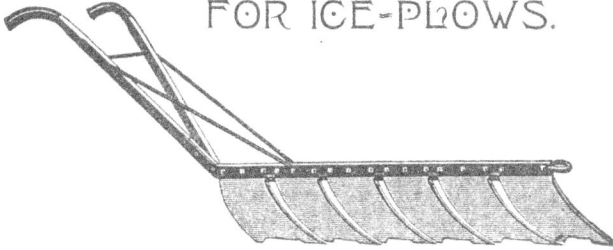

The Patent Clearing Tooth was invented in 1872, and consists of a flanged tooth in the forward end of the plow in place of the old-fashioned blank.

The benefits of this clearing tooth upon Ice Plows have become so universally appreciated that we cannot sell the old style beside them, and consequently we now make all our plows with this valuable improvement attached, placing upon the market

THE EASIEST WORKING PLOWS EVER MANUFACTURED.

All practical ice-cutters know that deep plows, with the old-fashioned blank tooth, are apt to run with difficulty through the chips which unavoidably get brushed into the grooves, especially if the chips are damp, as they will be on a "soft" day; and that sometimes the team is suddenly brought to a standstill by their accumulation.

This difficulty is in a measure overcome by the Clearing-Tooth, which, with its curved face, similar in width and shape to the other teeth, permits no chips to wedge in beside it or to block up before it, unless very damp, but neatly carries them out of the groove without any apparent resistance, thus leaving a clear groove in which the cutting teeth of the plow can freely perform their work.

With this brief description of its operation we think the advantage of it will be seen by every one, and we trust that those who contemplate buying plows will deem it to their interest to possess no other kind.

Clearing-Teeth can be easily inserted into old plows, and many cases have occurred where a badly working plow has been changed, by the addition of this improvement, into a good working one.

We are the sole purchasers of the patent.

❀ EVERY ICE DEALER ❀

should possess a sample, at least, of our plows, and prove by their practical working that they are all we claim for them.

Some of the Distinctive Points of our Ice Plows are:

Best cast steel in the teeth,—no iron teeth whatever.

A correct curve to the face of the teeth, insuring ease of cutting.

The Patent Clearing-Tooth applied to all sizes.

Stiff, substantial backs.

Extra large chip spaces cut in the backs or beams, enabling a plow to cut its Full Depth, a feature which is wanting in the plows of many manufacturers.

Raised head pieces on all plows excepting the deep ones.

Norway iron handle connections, where some use refined or malleable iron, which are very liable to break.

A general good proportion and finish, which secures not only strength, but a shapely appearance, and the easiest possible working qualities.

Every plow is furnished with a substantial case, riveted, ironed, painted, and strapped, which, besides protecting the teeth from injury, renders the moving of the plow from place to place on the ice an easy matter.

Filing Gauges furnished free to all customers who may send for them.

☞ *Order by this year's numbers.*

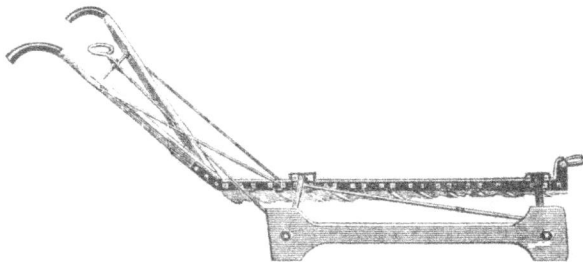

MARKER, WITH SWING GUIDE.
Weight complete, 146 lbs.

No. 1.—3¼in. CAST STEEL MARKERS, with Swing
Guide, complete . . . . each, $58.00

The marker is made with eleven cutting-teeth, and grooves to the depth of three inches, which amount is cut at once. Its use is to mark out the field of ice into squares, preparatory to grooving with the plows.

Having made a straight line by means of a hand-plow, or a line-marker, the marker-teeth should be run in this line across the field, in this manner making the first groove. The guide is then swung over the back of the marker, and returning follows in the groove cut, thus regulating the distance for the teeth to cut a second groove, parallel to the first.

Parties who have much ice to cut should use a marker instead of a swing-guide plow, as many try to do. The marker cuts a wider groove, cuts faster, always has the guide on, and, being shallow, is much more apt to mark straight than a deep plow with guide.

The regular width of our guide is twenty-two inches; other sizes made to order.

When two guides for one marker are wanted, in order that two-sized cakes can be cut, the prices for the second will be as follows:—

EXTRA GUIDES, 28 inches wide, or less . . . . each, $7.50

             30 ,,   ,, to 36 inches wide . . . . ,,   8.00

             40 ,,   ,, to 44 ,,   ,, with . .

             solid-forged double brace . . . . . ,,   10.00

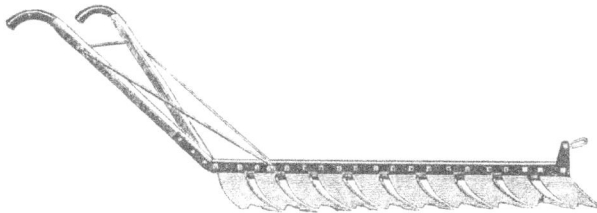

6-INCH PLOW—9 TEETH.

Weight, with case, 117 lbs.

No. 2.— 6-inch CAST STEEL PLOWS, 9 cutting-teeth, each   . . $48.00

 ,, 3.— 6-inch   ,,   ,,   7   ,,   ,, . . . 42.00

 ,, 4.— 7-inch   ..   ,,   7   ,,   ,, . . . 44.00

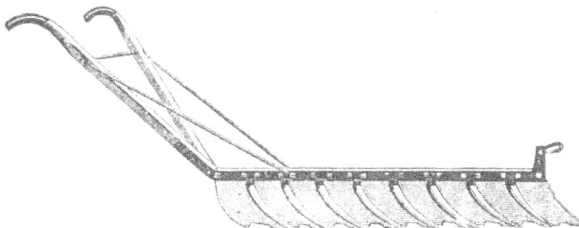

8-INCH PLOW—8 TEETH.

Weight, with case, 125 lbs.

No. 5.— 8-inch CAST STEEL PLOWS, 8 cutting-teeth . . each, $50.00

 ,, 6.— 8-inch   ,,   ,,   7   ,,   ,,   48.00

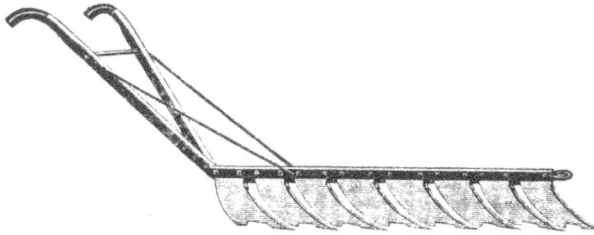

9-INCH PLOW—7 TEETH.
Weight, with case, 126 lbs.
No. 7.—9-inch CAST-STEEL PLOWS, 7 cutting-teeth, . . each, $51.00

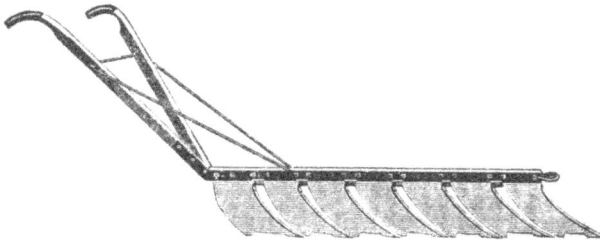

10-INCH PLOW—6 TEETH.
Weight, with case, 125 lbs.
No. 8.—10-inch CAST-STEEL PLOWS, 6 cutting-teeth, . . each, $53.00

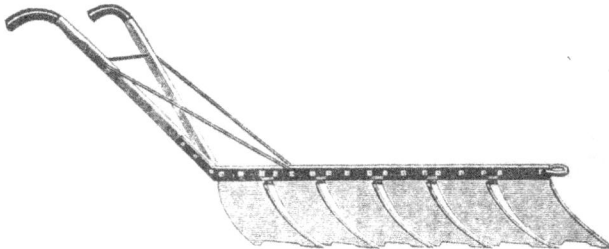

12-INCH PLOW—5 TEETH.
Weight, with case, 126 lbs.
No. 9.—12-inch CAST-STEEL PLOWS, 5 cutting-teeth, . . each, $58.00
„ 10.—14-inch   „   „   5   „   . . „   65.00

The use of the Ice Plow is to groove the ice to the depth required, by following in the marker grooves. The usual depth for grooving is about two-thirds the thickness of the ice; and the amount a plow will cut at once is ¼ inch, or a little more, to each tooth. Thus, a seven-toothed plow is capable of cutting 1¾ inches easily; a nine-toothed plow, 2¼ inches; and so on.

All our plows will groove to their FULL DEPTH; and customers, in buying, should order as shallow plows as will answer their purpose, as they are naturally stiffer, and having more teeth, will cut faster than deep ones.

☞ *It should be borne in mind that our plows are so graded that the marker cuts the widest groove and the deep ones the narrowest; and in no case should a shallow plow be run in a groove made by a deep one.*

NOTE.—All our plows have the Patent Clearing Tooth. See page 18.

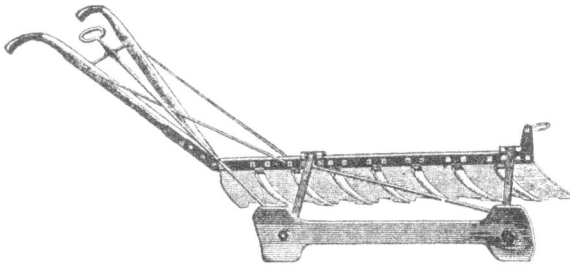

SWING GUIDE PLOW.

No. 11. — 6-inch CAST-STEEL PLOWS, 7 cutting-teeth, with swing
        guide attached ; weight, with case, 140 lbs.,   . . . each, $50.50
No. 12. — 7-inch CAST-STEEL PLOWS, 7 cutting-teeth, with swing
        guide attached ; weight, with case, 142 lbs.,   . . .   ,,   52.50
No. 13. — 8-inch CAST-STEEL PLOWS, 7 cutting-teeth, with swing
        guide ; weight, with case, 152 lbs., ...   ,,   56.50
No. 14. — 9-inch CAST-STEEL PLOWS, 7 cutting-teeth, with swing
        guide attached ; weight, with case, 162 lbs.,   . . .   ,,   59.50
No. 15. — 10-inch CAST-STEEL PLOWS, 6 cutting-teeth, with swing
        guide attached . . . .   ,,   61.50
No. 16. — 12-inch CAST-STEEL PLOWS, 5 cutting-teeth, with swing
        guide attached . . . .   ,,   66.50
No. 18. — SWING GUIDES, only, 22 inches, with connections .   ,,   8.50
    When two guides for one plow are wanted, in order that two-sized cakes
can be cut, the prices for the second guides will be as follows : —
EXTRA GUIDES. — 28 inches wide, or less . . . . each, $7.50
    ,,   ,,   30 inches wide, to 36 inches wide . . .   ,,   8.00
    It is not practicable to use guides wider than 36 inches on plows.

    A plow with swing guide attached is a device for combining a plow
and marker, and for dealers to put up small quantities of ice, is a cheap and
convenient tool to perform their work.

    The swing guide should be attached to the plow *only* for the purpose of
marking out the field of ice. (The grooves thus made will be about two
inches deep, instead of three, as in the case of markers.) When this is
done, it should be taken off, in order that the plow may be allowed to cut
to the required depth.

    Swing guides should not be attached to plows deeper than eight inches,
although nine inches is allowable, as on deeper ones they are not apt to
work as well. When it is necessary to cut as deep as ten or twelve inches,
we should advise the use of a shallower plow first (a seven or eight inch),
as a shallow one cuts faster and makes a wider groove, enabling the deep
one to follow much more easily.

    The use of the regular marker with separate plow is always to be recom-
mended in preference to the swing guide plow, where the capital of the
purchaser will allow.

No. 19.—STATIONARY GUIDES FOR PLOWS . . . . . . each, $5.50

Plows, with Stationary Guides attached, are but little used on account of their inconvenience, as they oblige the operator to go around the field instead of returning by the groove last cut.

We make them to order, only.

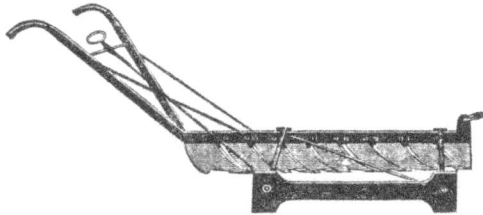

MAN OR PONY PLOW.
Weight, with case, 100 lbs., complete.

No. 20.—MAN OR PONY PLOWS, 6 inches, with 22 inch swing
guide . . . . each, $40.00

The Man Plow contains seven teeth, and clearing tooth, cuts a narrow groove six inches deep, is light to handle, and is capable of grooving about one inch in depth at a time, when drawn by a man, another being required to hold the handles. The Swing-Guide which is attached, is for marking out only, and should be detached when the plow is run deeper. Our regular width of Guides is 22 inches. Other sizes will be made to order.

HAND PLOW.
Weight of 5½ inch plow, with case, 18 lbs.

No. 22.—CAST-STEEL HAND PLOWS, 5½ inch . . . . . . each, $14.50
„ 23.   „   „   6   „ . . . . .   „   15.00

This tool is used principally where ice is housed in blocks containing two or more cakes, to groove between the cakes preparatory to splitting them apart with house bars when discharging. It is useful, also, in making the first line on the ice-field, and in finishing out the ends of the grooves. The usual size is 5½ inches.

We strap a painted case on every plow to protect the teeth.

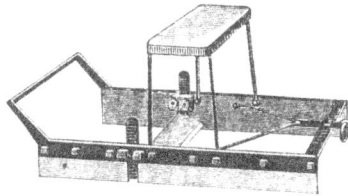

SNOW ICE PLANE.

Weight, with one knife, 135 lbs. ; with two knives, 155 lbs.

No. 24. — Snow Ice Planes, with one knife . . . . . . each, $38.00

 „   25. — Extra Knives for planes . . . . „   10.00

The invention of clamping the knife to the plane, instead of screwing it in, was made in our manufactory in 1869, and the change has proved so highly satisfactory that we make them now entirely of this improved pattern. The knife is held in this manner perfectly firm, while it can be taken out and exchanged much more quickly than when fastened by the old method.

We send with each plane a wrench for the nuts.

Persons who intend to do much planing should have two knives, so that while one is being ground the other can be in use.

Extra Knives will fit any of our planes made after this pattern, and can be sent separately at any time.

The use of this implement is to shave off snow ice, and in some seasons it is indispensable to procuring good ice.

If the grooves made in marking are 22 inches apart, the sides or guides of the plane will run on the bottom of these grooves, and the knife can be adjusted to cut any part of their depth.   By the use of two horses, three inches can be taken off at once.

Planing can be much more smoothly done by using a 21½-inch guide on the Marker, and by putting on a check-gauge, (see page 25) by which the marker is made to groove exactly the depth required to be planed off.   Then the Plane, which is 22 inches wide, by having its knife set to the bottom of the sides or guide plates, will lap over one half inch on to the planed portion, and cutting out the marked grooves completely, will leave the surface as smooth as new ice.   This method is practised by the most particular shippers, and is decidedly to be recommended to all who have to plane.

We make markers for this method of planing, with a half-inch collar on the guide bar next to the guide plate, which by being put on the outside of the plate, will cause the guide to be narrowed to 21½ inches.

Persons desiring to have their guides altered so as to cut both 21½ and 22 inches, in order that they may adopt this method of planing, can do so at a moderate expense by sending the guides to us.

Do not delay till the last moment.

(Please observe Shipping Directions on page 4.)

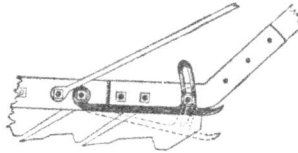

PLANING GAUGE.

No. 26. — PLANING GAUGES, complete with bolts . . . . each, $1.00
No. 27. — RUNNERS FOR PLANES, complete with bolts . per pair, $2.50

The Planing Gauge shown above is a neat device for making the marker cut exactly the desired depth of groove. When the plane runs wider than the marker-grooves, it is important that the grooves be no deeper than what is required to be taken off, as the plane must cut to the bottom of the grooves if the knife is set clear down, as is positively essential in this manner of planing.

Runners are only necessary when parties wish to plane after the plow grooving (which must be just 22 inches apart) is all done. They bolt on the inside of the plane and make a bearing for the plane to run on.

ICE SAWS.

Weight, with cases, 4 ft. 13½ lbs.; 4½ ft. 15 lbs.; 5 ft. 16½ lbs.

No. 28. — ICE SAWS, 4 feet, best quality, with case . . . . each, $5.00
„ 29. „ 4½ „ „ „ . . . . „ 5.25
„ 30. „ 5 „ „ „ . . . . „ 5.50
„ 31. „ 4 „ „ without case . . . . ., 4.50
„ 32. „ 4½ „ „ „ . . . . „ 4.75
„ 33. „ 5 „ „ „ . . . . „ 5.00

Each of above saws is furnished with a neat, substantial case, having leather strap.

Unless instructions are given to the contrary, saws with cases will be shipped on orders.

Most manufacturers sell their saws without cases. In ordering please state which way you wish them sent.

The saw is a very necessary article in cutting ice, as it must be used in opening the channel, and also in sawing the sheets or "floats" out of the field. Five feet is the best length for fast work where the depth of water will permit its use.

The handle which we put on our saw blades is worth three of those generally used, and the wood cannot turn on the spindle.

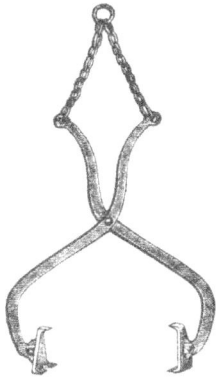

No. I.   No. 2.
HOISTING TONGS.
Weight, 20 lbs. each.

No. 35. — Hoisting Tongs, No. 1, with adjustable hands . each, $6.00
    36.   ,,   ,,   ,,   1,   ,,   extra heavy   .   ,,   7.00
    37.   ,,   ,,   ,,   2, with claw points . . .   ,,   4.50
    38.   ,,   ,,   ,,   2, ,,   ,,   ,, extra heavy   ,,   5.50

These are used principally for hoisting ice directly from the water to the platform.

Tongs No. 1 are made with joints, allowing the hands to adjust themselves to any reasonable shape or size of cake.   Our hands are made solid of steel, are very strong and durable, and of skeleton shape, rendering them less liable to slip on account of any unevenness of the ice.

Tongs No. 2 have no joint, being made with a stationary claw or double point, and will hold very firmly.

The Extra-Heavy sizes are made of 2x5-8 iron, are very strong, and are only used for hoisting or lowering the heaviest ice in double cakes.

LOWERING TONGS.
Weight, 6 lbs.

No. 39. — Lowering Tongs . . . . each, $2.00

These are really another variety of hoisting tongs.   They are made of heavier iron than hand-loading tongs, and are sufficiently strong for all ordinary hoisting.

The draught upon the shackle is sufficient to make them hold the ice perfectly secure, and they are especially adapted for places where there is a small amount of room.

For filling ice-boxes and market refrigerators there is nothing better.

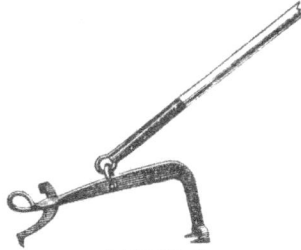

GRAPPLE.
Weight, 17½ lbs.; drop, 8 in.   Other sizes made to order.

No. 45. — GRAPPLES, with handles . . . . each, $4.25
„ 46.   „   without   „   . . . . „   3.75

The grapple, or drag, is used to draw cakes of ice up an inclined plane by horse or steam-power, in the absence of elevator machinery. A pole handle is attached, by which to carry it down the incline, though oftentimes the grapple is arranged to slide back in a trough or upon a wire, thus dispensing with the handle.

We make them of Norway iron, and they will do very severe service.

CANAL GRAPPLE.
Weight, 10 lbs.; drop, 7 in.

No. 47. — CANAL GRAPPLES, with rope, 12 feet . . . . . . each, $4.25

This grapple is used by some, instead of the floating-hook, for the purpose of drawing small floats of ice along the canal, especially where the distance is long.

It is not intended to be used for hoisting ice up the inclined plane, not being strong enough for that work.

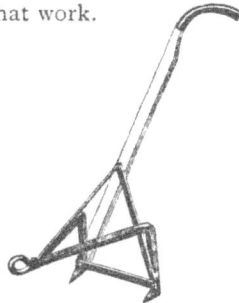

JACK GRAPPLE.
Weight, 20 lbs.; drop, 8 in.

No. 48. — JACK GRAPPLES . . . . each, $4.25

This kind of grapple is employed for the same purposes as No. 45. Being made with a stationary handle, it is considered better by some than the other pattern.   It is particularly liked in the West.

It is substantially made, with solid points, and steeled with the best steel.

CHISELS AND SMALL TOOLS.

None but the workmen who use small tools day after day know better than we do how much depends upon their shape, weight, and temper being exactly right.

Our long experience and careful attention have enabled us to bring them to apparent perfection.

We claim as distinctive points :

A large amount of best English steel.

A temper precisely adapted to the uses of each.

A general shape and weight which at once commend them to an experienced hand as capable of doing the most efficient service.

They are worth 50 per cent. more for real service than lighter and inferior tools.

BREAKING BAR.

Weight, 16 lbs.; length, 4 ft. 6 in.; pad, 7½ in. x 4¾ in.; chisel blade, 2⅝ in. wide.

No. 50. — Breaking Bars . . . . each, $3.25

This tool, having a wedge-shaped "pad," or blade, is used to insert in the plow grooves, and thus detach the sheets or floats from the main body of the ice, and also to break up these floats into strips. The chisel end is merely for convenience, in case a sharp end is required.

The pad end is provided with ears upon which to bear the foot, — an act which is necessary in detaching floats.

It has a blunt end, which is not intended to reach the bottom of the groove, the seam being wedged open.

HOUSE BAR.

Weight, 12 lbs.; length, 4 ft.; pad, 4 in. wide.

No. 51. — House Bars . . . . each, $2.50

The House Bar is like a breaking bar, excepting that it is made lighter, without any chisel end, and has a steel pad. When ice is stored in blocks containing two or more cakes, this bar is used to separate the cakes, and is made to fit the groove left by the hand plow.

It has no sharp edge.

No. 1.

Weight, 16 lbs.; length, 4 ft. 6 in.; fork, 8 in. wide.

No. 2.

Weight, 16 lbs.; length, 4 ft. 9 in.; fork, 8 in. wide.

No. 3.

Weight, 13 lbs.; length, 4 ft. 9 in.; fork, 6½ in. wide.

FORK BARS.

No. 52. — Fork Bars, No. 1	. . . .	each,	$4.25
53.	„	No. 2	. . . .	„	4.60
54.	„	No. 3	. . . .	„	4.00

Our fork bars are forged of solid steel, with chisel-shaped points, and are much superior to those made with prongs screwed in.

No. 1 is much preferred by many to the breaking bar for breaking off floats, and where the grooves have become more or less frozen, resulting from water having flowed into them, this bar is almost indispensable.

No. 2 is used in the same manner as No. 1, the only difference being a ring on the handle.

No. 3, having but three tines, is somewhat lighter than Nos. 1 or 2, and is, therefore, found more convenient for use on the floats or in the canal.

SPLITTING FORK, TWO TINES.

Weight, 16 lbs.; length, 4 ft. 9 in.; fork, 13½ in. long.

No. 60. — Splitting Forks . . . . each, $4.25

This tool differs from the ordinary Fork Bar in having but two tines, which are very long and thin. While the regular Fork Bar is intended to *wedge* off the ice without having the points strike the bottom of the grooves, as is the preferable way when the grooves are sufficiently deep, this bar is made to strike the bottom of the groove, and at the same time to wedge slightly. For thick ice, not deeply grooved, it is a capital tool. The tines are wholly steel.

CALKING BAR.

Weight, 12 lbs.; length, 4 ft. 9 in.; blade, 17 in. x 4¾ in.

No. 61. — CALKING BARS . . . . each, $2.50

This bar is used to calk the ends of the grooves on the field of ice, and on the floats before they are detached, with the chips made in grooving, in order to prevent the water from running in.

The blade is made long and thin, suitable for calking the deepest grooves ; and in order to insure stiffness, we make it entirely of steel.

BAR CHISEL.

Weight, 16 lbs.; length, 4 ft. , in.; blade, 12 in. x 4½ in.

No. 65. — BAR OR PACKING CHISELS . . . . each, $3.25

The Bar Chisel is made with a wide blade, bevelled on one side, and is used mostly to cut around the cakes in getting ice out of the house. It is also used for trimming off any unevenness of the blocks when stowing ice in the house, and for breaking out the canal, cutting holes, or any work requiring a rapid cutting chisel.

SOCKET CHISEL.

Weight, 9 lbs.; length, 4 ft. 9 in.; blade, 10½ x 5 in.

No. 66. — SOCKET CHISELS . . . . each, $3.25

This is made with a blade like a bar chisel, and with a wooden handle, making it lighter in weight, on which account some prefer it in packing.

It is much used in loading vessels, being convenient to use in close quarters.

FLOOR CHISEL.

Weight, 9 lbs.; length, 4 ft. 9 in.; blade, 10½ x 5 in.

No. 67. — FLOOR CHISELS . . . . each, $3.25

The Floor Chisel is especially made to use in vessels, when, in loading, it is necessary to shave a layer of ice so its surface will be smooth and level, thus permitting other cakes to slide over it without the use of runs.

This operation is called " shaving a floor."

It is equally as serviceable for this work in ice-houses.

STARTING CHISEL.
Weight, 14 lbs.; length, 4 ft. 6 in.; blade, 2½ x 10 in.

No. 68. — STARTING CHISELS . . . . each, $3.25

This chisel is made for the purpose of starting up the blocks of ice in the house, after they have been cut around with the bar chisel, and is curved to prevent too great stooping of the body.

We make them of very heavy steel in the bend and its approach, a weakness there being a fault too common in this kind of chisel.

This tool cannot be surpassed for "striking up." It is also the best bar for "wetting down" the snow, in the list.

SEPARATING CHISELS.
Weight, 14 lbs.; length, 4 ft. 6 in.; blade, 16 in. x 3¼ in.

No. 69. — SEPARATING CHISELS . . . . each, $2.75

This chisel is made for the purpose of separating cakes of ice in the ice-house *when packed on edge*, as is almost universally done in the West. Having a long, thin blade made of steel, it is the most perfect tool in use for that work.

The Starting Chisel is used with it for striking under the cakes. When ice is packed *flat*, the Bar Chisel should be used instead of the Separating Chisel.

SPLITTING CHISEL.
Weight, 13 lbs.; length, 4 ft. 6 in.; blade, 10½ x 3 in.

No. 70. — SPLITTING CHISELS . . . . each, $2.25

The Splitting Chisel is made with a narrow blade, and with a gradual bevel on both sides. It is used to split blocks or strips into cakes, either in the canal or on the platform, and also comes into use for various purposes, being a light and handy tool.

RING HANDLE CHISEL.
Weight, 13 lbs.; length, 4 ft. 9 in.; blade, 10½ x 3 in.

No. 71. — RING HANDLE CHISELS . . . . each, $2.50

This chisel is made the same as a splitting chisel, with the exception that it has a ring on the handle for the hand, and it is used for the same purpose. It is particularly useful in splitting strips in the canal, and in cutting holes on the pond when the ice is thin and it is desired to let the water up to saturate a fall of snow,—in which case the ring forms an effectual safeguard against losing the chisel.

CANAL CHISEL.
Weight, 15 lbs.; length, 6 ft.; blade, 10½ x 3 in.

No. 72. — Canal Chisels, No. 1, Iron Handle . . . . each, $2.50

„ 73. — „   „   No. 2, Wood Handle . . . „   2.50

This chisel is used for splitting cakes from the strip while floating up the canal, when the operator is standing upon a raised platform.

No. 1 is six feet in length, and has an iron handle, partly of 7-8 round iron, and is comparatively light.

No. 2 is also six feet long, and has a wooden handle, which some like better than the iron ones.

HOOK CHISEL.
Weight, 6½ lbs.; length, 6 ft. 9 in.; blade, 10½ x 2¾ in.

No. 74. — Hook Chisels . . . . each, $3.00

The Hook Chisel is a canal chisel and ice-hook combined, and has a wooden handle. It does not necessarily belong to a complement of tools, but is sometimes needed to draw ice into position to split. We make it heavy enough for splitting well, the hook being secondary in importance.

ICE BREAKER.
Weight, 5 lbs.; length, 4 ft. 8 in.; length of tines, 7½ in.; width at points, 2¾ in.

No. 75. — Ice Breakers, long handle . . . . each, $3.50

„ 76. — „   „   D handle . . . . „   3.50

This tool is used to break up ice into fine pieces for saloon or restaurant use.

Its square points are forged corner-wise, and as they are thrust down the sides of a cake of ice — which is put into a basket to hold the chips — they shave off pieces, coarse or fine, according to the pleasure of the user, in a rapid manner.

The Ice Breaker is now considered a necessary tool on city wagons which deliver considerable ice by the basket, and is very excellent for the use of fish packers and ice-cream makers.

We make them with both long and D handles.

ICE SHAVER.
Weight, 4 lbs.; length, with D handle, 3 ft. 6 in.; blade, 8 x 4¾ in.

No. 77. — Ice Shavers, D handle . . . . each, $2.25

„ 78. — „   „   long handle . . . . „   2.25

This is used for similar purposes as the Ice Breaker.

It is made in a first-class manner of best steel, and is in no sense a cheap tool, such as are quite commonly sold for breaking or shaving purposes.

ELEVATOR FORK.

Length, 6 ft. 7 in. ; tines, 6 in. long.

No. 80. — Elevator Forks . . . . each, $1.50

This article is deemed better than an ice hook for feeding ice to the elevator, inasmuch as two pushing points give a person better control of the cake than one.   The puller is placed on the opposite side from the prongs, so that by simply turning it over, either side can be used without the user changing his position.   The handle is six feet long.

Longer ones made to order.

SIEVE SHOVEL.

Weight, 6½ lbs.; size of shovel, 16½ x 14 in.

No. 84. — Sieve Shovels, 30 in. Handle . . . . each, $2.00
No. 85.   „   „   40 „   . , ...   „   2.00

This shovel or scoop is very useful in clearing the canal of chips and pieces of ice.   It is lighter than the scoop net, and some prefer it.

Both lengths of handle are used.   The shorter one is the same as an ordinary shovel, suiting some, while others want the additional 10 inches.

SCOOP NET.

Weight, 7 lbs.; handle, 5 ft.; hoop, 15 in. diam.

No. 86. — Scoop Nets . . . . each, $3.50

These are made as light as possible by the use of welded chain in place of jack chain, which, although a more expensive way, is much more serviceable.

The scoop net is used to remove from the water small pieces of ice which obstruct the channel more or less, and it has the advantage of not freezing so as to impair its straining qualities.

ICE AUGER.
Weight, 6 lbs.; length, 3 ft. 6 in.

No. 88. — ICE AUGERS, 1½ inch . . . . each, $3.50

This auger is made with an easy-working head, and a long shank, enabling the person to retain a standing position while using. It is used principally in making holes through which to measure the ice, as well as in which to insert pins for stretching lines, towing floats, and other purposes.

We have this year made an improvement in our augers, causing them to cut much more freely and rapidly than before.

They will give entire satisfaction.

Old ones can also be altered over.

MEASURING IRON.

No. 89. — MEASURING IRONS . . . . each, $0.50

The measuring iron has the end turned to catch on the bottom of the ice, and is marked off by inches as high as twenty-two.

It is polished throughout.

LINE MARKER.
Weight, 2 lbs.; length, 4 ft. 8 in.

No. 90.—LINE MARKERS . . . . each, $0.90

This tool is a simple contrivance for the purpose of making a line by means of a straight edge in which to run the marker teeth, thus making the first groove in marking out the field of ice. For those who have no hand plow, it is a sufficient substitute for this purpose.

HAND SAW.
Weight, 2½ pounds.

No. 92.—HAND SAWS, 30 in., Wood handle . . . . . . each, $1.50
No. 93.— ,, ,, ,, Iron ,, ..., 1.50

The hand saw is preferred by some retailers to the axe for wagon use. The axe, however, is much more generally used.

TAPPING AXE.
Weight, 5 lbs.

No. 94.—Tapping Axes, 2¾ in. wide . . . . each, $2.00

Some persons prefer the tapping axe to the ring chisel for the purpose of cutting holes through the ice, when it is necessary to wet down a field of ice.

This axe is also used by some in the ice-house to cut around the cakes of ice instead of using the bar chisel.

In both cases, however, we consider the chisel much the most suitable.

(Boston Pattern.)	**ICE AXES.**	(Chicago Pattern.)
Weight, 3½ lbs.		Weight, 3¾ lbs.

No. 95.—Ice Axes, 30 in. handle, Boston pattern, 1¾ in.
wide . . . . per doz., $18.00
No. 96.—Ice Axes, 30 in. handle, Chicago pattern, 2½ in.
wide . . . ., „ 18.00
No. 97.—Ice Axes, 30 in. handle, New York pattern, 3 in.
wide . . . . „ 18.00

The use of the axe is well known to be for splitting cakes into pieces suitable for retailing. The latest Eastern pattern has a narrow bit about 1¾ inches wide. The Chicago pattern has a 2½-inch bit. The New York pattern has a 3-inch bit. Our regular length of handle is 30 inches; other lengths to order.

A skilful workman, with a good narrow-bit axe, can split a cake of ice into small, shapely pieces, quicker, and with less waste, than by the saw or any other tool.

SOUTHERN PATTERN ICE AXE.	MILKMEN'S HATCHET.	CHEST HATCHET.
Weight, 3 lbs.	Weight, 2 lbs.	Weight, 1 lb.

No. 98. — Ice Axes, 20 in. handle, Southern pattern, 2½ in.
wide . . . . per doz., $18.00

This pattern of ice axe suits in localities where the dealers have become accustomed to a short-handled axe.

No. 100. — MILKMEN'S AXES, 18 in. handle, 2 in. wide .   per doz., $15.00

Designed to meet the wants of those who have occasion to subdivide the cakes left by the ice-man, and who wish an axe of convenient size and length of handle to work inside of an ice-box.

No. 102. — CHEST HATCHETS, 13 in. handle, 1½ in. wide, per doz., $10.50

This neat and handy article is intended for ice-chest purposes, and is very convenient to split off small pieces of ice for table use.

It is made of as good iron and steel as a larger axe, and is not in any respect a cheap article.

AMERICAN ICE-CHISEL.	STAR ICE CHIPPER.	ICE AWL.
(WOOD HANDLE.)		
Length, 10 in.; width, 2½ in.	Length, 10 in.; weight, 1¼ lbs.	Length, 9 in.

No. 104. — AMERICAN ICE CHISELS, wood handle .   .   .   per doz., $3.25

This convenient tool is substantially made and of good size for ice-chest use, capable of shaving ice very rapidly. It has a full polished blade, solidly fastened in, and the shank does not go through the handle, thus causing the wood to be perfectly smooth to the hand.

No. 105. — AMERICAN ICE CHISELS, iron handle   .   .   .   per doz., $2.50

This style of American Ice Chisel is a little smaller than the wood-handled one, has a black blade, and a solid iron handle.   Many prefer it on account of its convenience in cracking a piece of ice while being held in the hand.

No. 106. — STAR ICE CHIPPERS .   .   .   .   each, 30 cts. ; per doz., $3.00

A new chipper or shaver, with the advantage of having a gauge on it to prevent getting the shavings too coarse.   It is an excellent tool for ice-cream use on this account.

No. 107. — ICE AWLS (spear point)   .   .   .   each, 20 cts. ; per doz., $2.25

This is the best article for picking pieces of ice off a cake for ice-water, or butter, or for similar purposes in use.   It does not waste the ice as much as the chisels, but it is not suitable for shaving ice.

No. 108. — ICEMEN'S AWLS (round points) . each, 25 cts. ; per doz. $3.00

Used by some as a substitute for the axe or saw.

ICE HOOKS.

Our Ice Hooks are manufactured of the best **Norway** iron, and are heavily steeled, making them very stiff and strong, without being heavy and unhandy. The practical test will prove them second to none in the market.

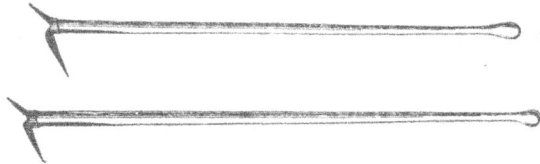

ICE HOOKS.

Weight of 4½ ft. Ice Hooks, per dozen, 37 lbs.

No. 110.—ICE HOOKS, 3 and 3½ feet, turned ash hdls., per doz., $9.25

,,	4 and 4½	,,	,,	,,	,,	9.50
,,	5	,,	,,	,,	,,	10.00
,,	6	,,	,,	,,	,,	10.50
,,	8	,,	,,	,,	,,	12.50
,,	10	,,	,,	,,	,,	14.00

Floating Hooks, shaved spruce, or turned ash handles, 12 to 18 ft. . . . . ,, 15.00

The Ice Hook, though so simple in its appearance, is absolutely necessary in the handling of ice, and is to every ice-dealer a very important part of a " set of tools."

FLOATING HOOKS have handles from 12 to 18 feet long, and are used to float the large sheets of ice from the "field" to the elevator. We have a large stock of shaved spruce poles for these hooks constantly on hand ; also, pullers and shovers for repairing.

When no length of handle is mentioned by parties ordering hooks, 4½ feet is understood, that being the common length.

Where ice is thick, 4 ft. handles are generally preferred. 3 and 3½ ft. handles are used principally in the holds of vessels.

☞ *We have a full line of best Ash Handles constantly on hand at lowest prices.*

CAR ICE HOOK,

No. 112.—CAR ICE HOOKS, 4 ft. . . . . per doz., $12.00

This article is the same as the common ice hook, excepting that the shover iron is carried down on the handle some eighteen inches in length, in order to strengthen and protect the wood for car use. The usual length is 4 ft., although 3 ft. 8 in. is a length used by a good many car-packers.

PATENT CANT ICE HOOK.

No. 115.—PATENT CANT ICE HOOKS, combined with 4 or 4½
       foot Ice Hooks (as shown in cut), · · · per doz., $16.00

No. 116.—PATENT CANT ICE HOOKS, unattached · · ·   „   6.00

The use of the Cant Ice Hook is to "edge up" ice when packing into houses or barges, as is the custom with most Western packers.

When no directions are given in ordering, we shall send 4-foot handles and right-handed hooks (like engraving), being the most common arrangement.

No. 120.—PULLERS, for Ice Hooks . . . . per doz., $4.50

No. 121.—SHOVERS   „   „   . . . .   „   2.75

No. 122.—RINGS   „   „   . . . .   „   .35

No. 123.—RIVETS   „   „ (2 doz. required for 1 doz. hooks) „   .06

ICE TONGS.

We give herewith engravings of ice tongs of the Boston, New York, Philadelphia, and Cincinnati patterns.

The Boston style is used throughout New England and in most of the Western States, and we consider it the most desirable shape for peddling use.

We are making all our tongs now of STEEL, finding that the demand is for stiff, light, handy tongs. We consider them the best models of what the ice-men desire, that we have ever produced, and we hope to place them in the hands of all deliverers.

The bows are made of good firm steel, and Best Cast Steel is welded into the points.

The favorite Swell Handle is used on all our tongs.

The price of our tongs has to be based upon their quality. The goods cannot be compared with the cheap ones ordinarily sold.

They are made throughout upon honor, for service, and for a reputation.

	13 in.	14½ in.	16½ in.	24 in.
Wts.:	2 lbs.	2¼ lbs.	3 lbs.	4¼ lbs.

STEEL TONGS (Boston Pattern).

No. 125. *Steel Tongs, family size, span 13 in., swell handle, per doz., $13.00
" " small " " 14½ " " " " 14.00
" " medium " " 16½ " " " " 15.00
" " loading " " 24 " " " " 17.00

HOLLOW HANDLE TONGS.

No. 127. †Hollow Handle Tongs, family size, span 13 in.,
Best Cast Steel ; weight, 2 lbs. . . . . . per doz., $21.00
Hollow Handle Tongs, small size, span 14½
in., Best Cast Steel ; weight, 2¾ lbs. . . . " 24.00
Hollow Handle Tongs, medium size, span 16
in., Best Cast Steel ; weight, 3 lbs. . . . " 24.00

The handles of these tongs are made of gas-pipe ; size is thus secured without making them top-heavy. The 13 in. tongs have handles ⅝ inch in diameter, and the 14½ and 16 in. tongs ⅞ inch in diameter.

The bows as well as the points are made of Best Cast Steel.

Ice men who once use these tongs will never give them up for any other style.

NEW YORK TONGS.   **PHILADELPHIA SEEL TONGS.**

No. 128. New York Tongs. Price same as Boston Tongs. Made of Steel.
No. 129. Philadelphia Steel Tongs, small size . . . . per doz., $15.00
" " " medium size . . . " 18.00
" " " large size . . . . " 21.00
" " " heavy loading size . " 24.00

* Steel Tongs are made of ordinary steel, with a fine steel in points. Stamped "Steel."
† These Tongs are all "Best Cast Steel," and so stamped.

CHAIN TONGS.
(CINCINNATI PATTERN.)
Weight: 2 lbs.

DRAG OR STOWING TONGS.
8 lbs.

EDGING-UP TONGS.
4 lbs.

No. 130. — CHAIN TONGS, 10 in. span . . . . per doz., $13.00
,, ,, 12 ,, ,, ... ,, .. 14.00
,, ,, 15 ,, ,, ... ,, .. 15.00

This style of tongs originated in Cincinnati, and is now getting to be used somewhat generally in Western cities. They are made entirely of steel, and are very light.

No. 135. — DRAG OR STOWING TONGS . . . . per doz., $21.00

These are used in some localities for stowing ice in houses; and they are made with long handles to enable the user to maintain a convenient position.

No. 136. — EDGING–UP TONGS . . . . per doz., $18.00

These tongs are very largely used in the Western country for edging up their ice when stowing in the house.

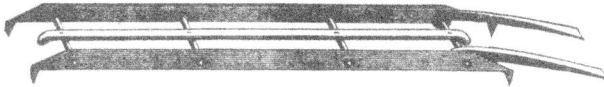

HOUSE RUN.
Weight, 6 ft. run, 76 lbs.

No. 140. — HOUSE RUNS, 6 feet . . . . each, $9.00

The house run is used in the houses and for loading wagons, and is made with a pair of flippers on the receiving end, as shown in the engraving, to cause a gradual rise of the ice.

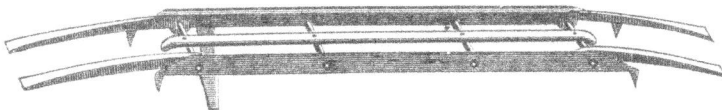

CAR RUN.
Weight, 7 ft. run, 105 lbs.

No. 141. — CAR RUNS, 7 feet . . . . each, $12.50

The car run is used for loading cars, and has flippers on the delivering as well as on the receiving end.

Two arms on the bottom are to prevent the run from being jarred ahead when receiving the ice.

SNOW SHOVEL.

DUNNAGE FORK.

No. 144.—Snow Shovels . . . . each, $1.25

This is used for handling snow, sawdust, etc. It is No. 7 size, made of steel, not polished.

No. 145.—Dunnage Forks, 4 tined . . . . each, $0.75

„ 146.— „ „ 5 „ . . . . „ 1.00

Used for moving packing material about ice-house.

PLOW ROPE.
Length, 9 ft.

No. 147.—Plow Ropes . . . . each, $1.50

These are made of the best manila 3-inch rope, with a thimble spliced in one end for the whiffletree hook, and a pair of sister-hooks in the other end by which to fasten to the plow.

Made with sister-hooks in both ends to order.

WAGON SCALES.

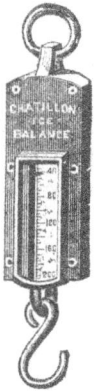

The "Ironclad" scales are made in a very strong manner for the especial purpose of weighing ice, and give universal satisfaction for a low-priced scale.

The "Straight" Scales are preferred by those who wish to weigh more accurately than by 5 pound marks.

The "Tubular" Ice Scales are far superior to any other kind of spring scales, and while they are expensive, they are worth their price for durability and reliability.

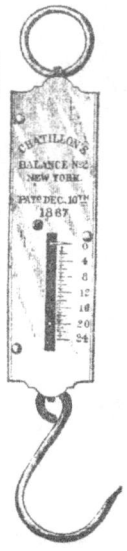

No. 150.—Ironclad Ice Scales, 200 lbs., by 5 lbs. $3.00

No. 160.—Ironclad Ice Scales, 300 lbs., by 5 lbs. 3.50

No. 170.—Ironclad Ice Scales, 400 lbs., by 5 lbs. 4.00

Ironclad Scale.

No. 172.—Straight Ice Scales, 100 lbs., by 1 lb. . . . 3.00

No. 173.— „ „ „ 150 „ „ 1 „ . . . 4.00

No. 174.— „ „ „ 200 „ „ 2 „ . . . 4.50

No. 176.—Tubular „ „ 200 „ „ 2 „ . . . 8.00

No. 177.— „ „ „ 300 „ „ 2 „ . . . 9.00

Straight Scale.

HOISTING TOP GIN.

WHARF GIN.

No. 180.—Top Gin, 9 inches, weight, 20 lbs. . . . . each, $4.25

    ,,   ,, 12 ,,   ,   27 ,, ... ,,   5.25

    ,,   ,, 14 ,,   ,   35 ,, ... ,,   6.50

No. 181.—Wharf Gins, 9 inches, weight, 24 lbs. . . . . ,,   5.00

    ,,   ,, 12 ,,   ,,   31 ,, ... ,,   6.00

    ,,   ,, 14 ,,   ,,   40 ,, ... ,,   7.25

These gins are strongly made, all of iron, and are generally preferred to wooden blocks.

SELF·LUBRICATING GINS.

These Gins can be run at quick speed without the use of any oil, being self-lubricating and are always ready for use; advantages which will be readily admitted by the Ice Trade, where, through the carelessness of employees, the Gins are often neglected to be oiled and soon wear out, whereas, our improved Self-lubricating Gins are always ready for service.

No. 182.—

Upper Self-lubricating Gins . . . 8-in. $5.00

Upper Self-lubricating Gins . . . 10-in. 6.00

Upper Self-lubricating Gins . , . 12-in. 7.50

Upper Self-lubricating Gins . . . 14-in. 9.00

Upper Self-lubricating Gins . . . 16-in. 10.50

SELF-LUBRICATING GIN.

No. 183.—Lower Self-lubricating Gins 8-in. $5.75

  ,,   ,,   ,,   ,,   . . . 10-in. 6.75

  ,,   ,,   ,,   ,,   . . . 12-in. 8.25

  ,,   ,,   ,,   ,, ... 14-in. 10.00

  ,,   ,,   ,,   ,, ... 16-in. 11.50

LANTERNS.

We sell the Genuine Tubular only—no imitations.

No. 185.—Patent Tubular Safety Lift Lanterns, per doz. $9.00

No. 186.— ,,   ,,   Regular   ,,   ,,   ,,   8.00

PATENT TUBULAR SAFETY LIFT LANTERN.

DUNNAGE BARROW.

These are convenient for removing shavings, sawdust, etc.

The barrow above illustrated, is 3 feet, 6 inches long, by 3 feet, 2 inches wide, and is 2 feet deep behind, and 1 foot, 8 inches deep in front.

No. 188.—DUNNAGE BARROWS . . . . each, $5.00

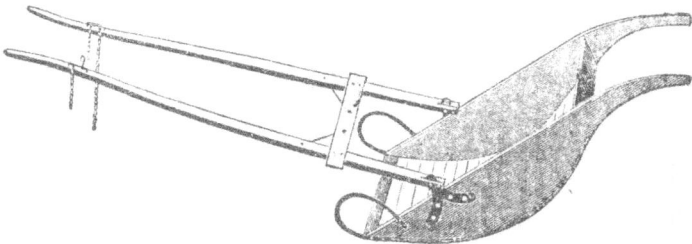

SCOOP SCRAPER.

No. 190.—SCOOP SCRAPERS . . . . each, $11.00
Shafts for same . . . . ,, 5.00

The scoop scraper is used for removing heavy snows, and all collections of snow made by the clearing scraper.

Its width is 3 feet, and it is made in every way in the most substantial manner, having a welded nose-piece on the front edge, and being otherwise well ironed.

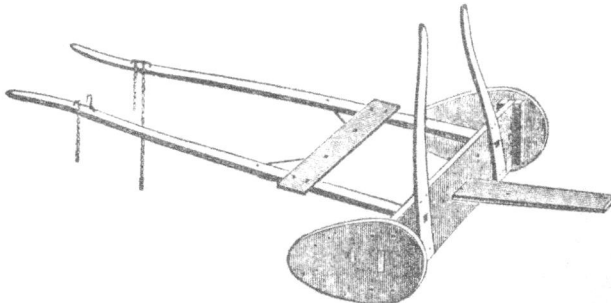

CLEARING OFF SCRAPER.

No. 191.—Clearing-off Scrapers, 6 feet long . . . . . . each, $9.00

,,   ,,   ,,   7 ,, ,, ...   ,,   9.25

..   ,,   ,,   8 ,, ,, ...   ,,   9.50

Shafts for same . . . .   ,,   5.00

The clearing-off scraper is used for scraping off light snows, and for cleaning up after the scoop scraper.

The usual length is 6 feet.

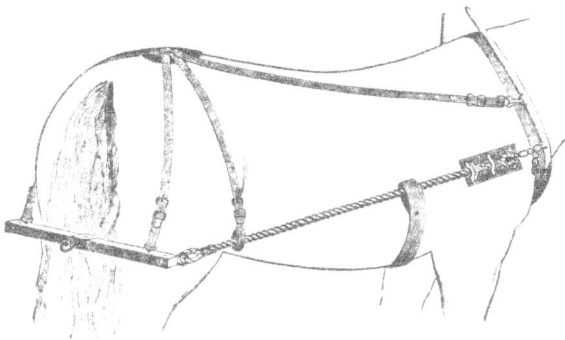

GROOVING HARNESS.

No. 195.—Grooving Harnesses . . . . each, $15.00
No. 196.—Hames, No. 10, strapped . . . . per pair, 2.50

This harness is especially designed for use in grooving, and is so made as to hold the whiffletree in the most advantageous position for receiving the end of the plow rope.

We make them of the best stock, and they will be found very serviceable, as well as convenient.

PATENT RUN IRON (New Pattern).

Weight per foot, 1 1-10 lbs.

No. 200.—Patent Run Iron, drilled, countersunk, and in 10, 12,

14, 15 or 16 feet lengths . . . . per foot, 5½ c.

Special figures for large lots of Patent Run Iron. A large stock constantly on hand.

This is the name given to a V-shaped iron, made especially to fasten on wooden runs. On account of the sharp surface presented, the ice will settle on the iron and be guided by it, the runs needing no sides to prevent the cakes from sliding off.

Our new pattern is 5⅝ in. deep, from top of V to the bottom of the bar. 14 feet is the usual length.

CHANDLER'S ICE CUTTER.

LOW'S ICE CRUSHER.

No. 205.—Low's Ice Crusher.

"Largest Size," for steam power; floor space, 4 feet square; weight,
    550 pounds $120.00
"Common Size," for steam or hand power; floor space, 3 feet square;
    weight, 400 pounds; capacity 2 to 3 tons per hour 65.00
"Common Size," for hand power; space, weight, and capacity nearly
    to above 60.00
"Medium Size," for hand power; floor space, 2 feet 3 in. square;
    weight, 220 pounds; capacity, 1½ tons per hour 40.00

No. 206.—Chandler's Ice Cutting Machines, viz. :—

No. 1, Japanned or Galvanized; small family use 4.00
No. 2, Japanned or Galvanized, with pan; large family use 6.00
No. 2½, Japanned, with pan; hotel use 10.00
No. 3, will take 25 pound piece of ice 25.00

It is a small, compact, simple, strong, and cheap machine, which crushes ice
with the *utmost ease and rapidity*. It has no equal. The No. 1 machine, as
shown in the above cut, occupies a space of only eight inches square, is about
twelve inches high, and can be attached by screws to a bar, counter, table, or shelf,
as desired. It turns easily, and can be operated by a child ten years old.

No. 210.—Patent Clearing Teeth.

Prices on page 48. See description page 18.

☞ In order to enable us to do a perfect job that can be warranted, plows needing Clearing Teeth should be sent to the factory, and we will straighten and file
them, ready for use, without extra charge above the clearing tooth price.

New cases are usually necessary when Clearing Teeth are put into old plows,
on account of the increased length.

No. 220.—Cases for Plows, markers to 7 in. inclusive . . each, $1.00
      ,,   ,,   ,,   8 in. to 12 in.   ,,   . . ,,   1.25

These are made of good pine lumber, riveted together, shod with a heavy strip of
iron screwed on, painted with two coats, and have two leather straps which buckle
over the beam of the plow.

☞ *All new Plows are furnished with cases free.*

Send for our Illustrated Article on Ice Harvesting, mailed free.

WM. T. WOOD & CO.,
49 North Market Street,
Boston, Mass.

PRICE LIST FOR 1888=89.

WILLIAM T. WOOD & CO.,

ICE TOOL MANUFACTURERS,

No. 49 North Market Street - - Boston, Mass.

FIRST-CLASS GOODS ONLY MANUFACTURED. ALL TOOLS WARRANTED.

Our prices are for full weight, long steel, fine quality tools only.

No.								Price.
1.	Cast Steel Markers, 3¼-in., with 22-in. Swing Guide, complete						each,	$58.00
	Extra Guides for Markers, 28-in. or less						,,	7.50
	,, ,, ,, 30-in. to 36-in.						,,	8.00
	,, ,, ,, 40-in. to 44-in., double braced						,,	10.00
2.	*6 in. Cast Steel Plows, 9 cutting-teeth						,,	48.00
3.	*6-in. ,, ,, 7 ,,						,,	42.00
4.	*7-in. ,, ,, 7 ,,						,,	44.00
5.	*8-in. ,, ,, 8 ,,						,,	50.00
6.	*8-in. ,, ,, 7 ,,						,,	48.00
7.	*9 in. ,, ,, 7 ,,						,,	51.00
8.	*10-in. ,, ,, 6 ,,						,,	53.00
9.	*12-in. ,, ,, 5 ,,						,,	58.00
10.	*14-in. ,, ,, 5 ,,						,,	65.00
11.	6-in. ,, ,, 7 ,, with 22 in. Swing Guide attached,						,,	50.50
12.	7-in. ,, ,, 7 ,, ,, ,, ,, ,, ,,						,,	52.50
13.	8-in. ,, ,, 7 ,, ,, ,, ,, ,, ,,						,,	56.50
14.	9-in. ,, ,, 7 ,, ,, ,, ,, ,, ,,						,,	59.50
15.	10-in. ,, ,, 6 ,, ,, ,, ,, ,, ,,						,,	61.50
16.	12-in. ,, ,, 5 ,, ,, ,, ,, ,, ,,						,,	66.50
18.	Swing Guides, only, 22-in., with connections						,,	8.50
19.	Stationary ,, only, 22-in.						,,	5.50
20.	Man or Pony Plows, 6-in., with 22-in. Swing Guide						,,	40.00
22.	Hand Plows, 5½-in.						,,	14.50
23.	,, ,, 6-in.						,,	15.00
24.	Snow Ice-Planes, 22-in. wide						,,	38.00
25.	Extra Knives for Planes						,,	10.00
26.	Planing Gauges, complete with bolts						,,	1.00
27.	Runners for Planes, ,,						per pair,	2.50
28.	Ice Saws, 4-foot, *best quality, with case*						each,	5.00
29.	,, ,, 4½-foot, ,, ,,						,,	5.25
30.	,, ,, 5-foot, ,, ,, ,,						,,	5.50
31.	Ice Saws, 4-foot *best quality*, without case						,,	4.50
32.	,, ,, 4½ ,, ,, ,, ,,						,,	4.75
33.	,, ,, 5 ,, ,, ,, ,,						,,	5.00
35.	Hoisting Tongs, No. 1, with adjustable hands						,,	6.00
36.	,, ,, No. 1, ,, ,, ,, extra heavy						,,	7.00
37.	,, ,, No. 2, with claw points						,,	4.50
38.	,, ,, No. 2, ,, ,, extra heavy						,,	5.50
39.	Lowering Tongs						,,	2.00
45.	Grapples, with handles						,,	4.25
46.	,, without handles						,,	3.75
47.	Canal Grapples, with rope, 12 feet						,,	4.25
48.	Jack Grapples						,,	4.25
50.	Breaking Bars						,,	3.25
51.	House Bars						,,	2.50
52.	Fork Bars, No. 1						,,	4.25
53.	,, No. 2						,,	4.60
54.	,, No. 3						,,	4.00
60.	Splitting Forks, 2 tines						,,	4.25

*It should be borne in mind that parties purchasing our Plows get the advantage of the valuable clearing-tooth without extra charge.

No.			Price.
61.	Calking Bars	each,	2.50
65.	Bar or Packing Chisels	„	3.25
66.	Socket Chisels	„	3.25
67.	Floor Chisels	„	3.25
68.	Starting Chisels	„	3.25
69.	Separating Chisels	„	2.75
70.	Splitting Chisels	„	2.25
71.	Ring-handled Chisels	„	2.50
72.	Canal Chisels, No. 1, iron handle	„	2.50
73.	„ „ No. 2, wood handle	„	2.50
74.	Hook Chisels	„	3.00
75.	Ice Breakers, long handle	„	3.50
76.	„ „ D handle	„	3.50
77.	Ice Shavers, D handle	„	2.25
78.	„ „ long handle	„	2.25
80.	Elevator Forks	„	1.50
84.	Sieve Shovels or Scoops, 30-in. handle	„	2.00
85.	„ „ „ 40-in. „	„	2.00
86.	Scoop Nets	„	3.50
88.	Ice Augers, 1½-in.	„	3.50
89.	Measuring Irons	„	.50
90.	Line Markers	„	.90
92.	Hand Saws, 30-inch, wood handle	„	1.50
93.	„ „ iron handle	„	1.50
94.	Tapping Axes	„	2.00
95.	Ice Axes, 30-in. handle, Boston pattern, 1¾-in. bit	per doz.	18.00
96.	„ „ „ „ Chicago „ 2½ „	„	18.00
97.	„ „ „ „ New York „ 3 „	„	18.00
98.	„ „ 20-in. „ Southern „ 2½ „	„	18.00
100.	Milkmen's Hatchets, 18-in. handle 2 „	„	15.00
102.	Chest Hatchets, 13-in. handle 1½	„	10.50
104.	American Ice Chisels, wood handle	„	3.25
105.	„ „ iron handle	„	2.50
106.	Star Ice Chippers each, 30 cts.	„	3.00
107.	Ice Awls, (spear point) „ 20 cts.	„	2.25
108.	Icemen's Awls, (round point) „ 25 cts.	„	3.00
110.	Ice Hooks, 3 and 3½ feet, turned ash handles	„	9.25
	„ 4 and 4½ „ „ „	„	9.50
	„ 5 „ „ „	„	10.00
	„ 6 „ „ „	„	10.50
	„ 8 „ „ „	„	12.50
	„ 10 „ „ „	„	14.00
	Floating Hooks, shaved spruce or turned ash handles, 12 to 18 feet	„	15.00
112.	Car Ice Hooks, 4 feet	„	12.00
115.	Patent Cant Ice Hooks, combined with 4 or 4½ feet Ice Hooks, right or left handed	„	16.00
116.	Patent Cant Ice Hooks, unattached	„	6.00
120.	Pullers for Ice Hooks	„	4.50
121.	Shovers for Ice Hooks	„	2.75
122.	Rings „	„	.35
123.	Rivets for Ice Hooks (2 doz. required for 1 doz. hooks)	„	.06
125.	Boston Hand Tongs, steel, 13-in. span, family size	„	13.00
	„ „ „ „ 14½ „ small „	„	14.00
	„ „ „ „ 16½ „ medium „	„	15.00
	„ „ „ „ 24 „ loading „	„	17.00
127.	Hollow Handle Tongs, best cast steel entire, 13-in. span	„	21.00
	„ „ „ „ „ „ 14½ & 16 "	„	24.00
128.	New York Tongs, same price as Boston Tongs		
129.	Philadelphia Steel Tongs, small size	„	15.00
	„ „ „ medium size	„	18.00
	„ „ „ large size	„	21.00
	„ „ „ heavy loading size	„	24.00
130.	Chain Tongs, 10-in. span, Cincinnati pattern	„	13.00
	„ „ 12 „ „ „	„	14.00
	„ „ 15 „ „ „	„	15.00
135.	Drag Tongs	„	21.00

No.		Price.
136.	Edging-up Tongs . . . . per doz.	18.00
140.	House Runs, 6 feet long (iron). . . . . each,	9.00
141.	Car Runs, 7 feet long (iron). . . . . ,,	12.50
144.	Snow Shovels, steel, black, No. 7 . . . . ,,	1.25
145.	Dunnage Forks, 4 tined . . . . ,,	.75
146.	,, ,, 5 ,, ... ,,	1.00
147.	Plow Ropes, with sister-hooks . . . . ,,	1.50
150.	Chatillon's Iron Clad Ice Scales, 200 pounds by 5 pounds . . . ,,	3.00
160.	,, ,, ,, 300 ,, ,, 5 ,, . . . ,,	3.50
170.	,, ,, ,, 400 ,, ,, 5 ,, . . . ,,	4.00
172.	,, Straight ,, 100 ,, ,, 1 ,, . . . ,,	3.00
173.	,, ,, ,, 150 ,, ,, 1 ,, . . . ,,	4.00
174.	,, ,, ,, 200 ,, ,, 2 ,, . . . ,,	4.50
176.	,, Tubular ,, 200 ,, ,, 2 ,, . . . ,,	8.00
177.	,, ,, ,, 300 ,, ,, 2 ,, . . . ,,	9.00
180.	Hoisting Top-Gins, 9-in., all iron . . . . ,,	4.25
	,, ,, 12 ,, ,, ... ,,	5.25
	,, ,, 14 ,, ,, ... ,,	6.50
181.	Wharf-Gins, 9-in., all iron . . . . ,,	5.00
	,, 12 ,, ... ,,	6.00
	,, 14 ,, ... ,,	7.25
182.	Upper Self-Lubricating Gins, 8-in. . . . . ,,	5.00
	,, ,, ,, 10 ,, ... ,,	6.00
	,, ,, ,, 12 ,, ... ,,	7.50
	,, ,, ,, 14 ,, ... ,,	9.00
	,, ,, ,, 16 ,, ... ,,	10.50
183.	Lower ,, ,, 8 ,, ... ,,	5.75
	,, ,, ,, 10 ,, ... ,,	6.75
	,, ,, ,, 12 ,, ... ,,	8.25
	,, ,, ,, 14 ,, ... ,,	10.00
	,, ,, ,, 16 ,, ... ,,	11.50
185.	Patent Tubular Safety-Lift Lanterns. . . . . per doz.	9.00
186.	,, ,, Regular ,, ... ,,	8.00
188.	Dunnage Barrows . . . . each,	5.00
190.	Scoop Scrapers, extra ironed, 3 ft. wide . . . . ,,	11.00
	Shafts for same . . . . ,,	5.00
191.	Clearing-off Scrapers, 6 ft. long . . . . ,,	9.00
	,, ,, 7 ,, ... ,,	9.25
	,, ,, 8 ,, ... ,,	9.50
	Shafts for same . . . . ,,	5.00
195.	Grooving Harnesses . . . . ,,	15.00
196.	Hames, No. 10, strapped. . . . . per pair,	2.50
200.	*Patent Run Iron, drilled, countersunk, and in 10, 12, 14, 15, and 16 feet lengths. Our new pattern . . . . per ft.,	.05½
205.	Low's Ice Crushers — "Largest Size," for steam power . . . . . . each,	120.00
	,, ,, "Common Size," for steam or hand power . . ,,	65.00
	,, ,, "Common Size," for hand power . . . . . ,,	60.00
	,, ,, "Medium Size," for hand power . . . . . . ,,	40.00
206.	Chandler's Ice-Cutting Machines, No. 1, jap'd or gal'd . . . . . ,,	4.00
	,, ,, ,, ,, 2, ,, ... ,,	6.00
	,, ,, ,, ,, 2½, ,, ... ,,	10.00
	,, ,, ,, ,, 3, ,, ... ,,	25.00
210.	Patent Clearing Teeth for Plows, 8-in. . . . . ,,	4.00
	,, ,, ,, 9 ,, ... ,,	4.50
	,, ,, ,, 10 ,, ... ,,	5.00
	,, ,, ,, 12 ,, ... ,,	6.00
	,, ,, ,, 14 ,, ... ,,	7.00

Straightening and filing included in above price when plows to have clearing teeth inserted are sent to factory.

220.	Cases for Plows, marker to 7-in., inclusive . . . . each,	$1.00
	,, ,, 8-in. to 12-in., ,, ... ,,	1.25

All new plows are provided with cases without extra charge.

WM. T. WOOD & CO.,

49 No. Market Street, Boston, Mass.

MUSEUM of AMERICAN ICE HARVESTS and WOODWORKING CRAFTS

This catalog was furnished for reproduction by the Museum o
American Ice Harvests and Woodworking Crafts, which is unde
development in a Milwaukee suburb. Correspondence addres
is 11458 North Laguna Drive 21W, Mequon, Wisconsin 53092
phone (414) 242 1571. We have over 250 ice tools from 2
states, the only substantial collection of natural ice day
artifacts in America. Among 26 (ice business) horsedraw
implements, is an ice wagon recently acquired from Oklahoma
This provided a pleasant surprise when the nameplates showe
it to have been made at nearby Oshkosh, Wisconsin.

To be able to display at least one example of each of the 9(
different tools used in natural ice harvesting, we need th
following:

> Covered ice wagon (p.9); 12" & 14" plow (p.21); canal
> grapple (p.27); breaking bar (p.27); house bar (p.28);
> separating chisel (p.31); canal chisel (p.32); eleva-
> tor fork (p.33); line marker (p.34); house, car runs
> (p.40); scoop scraper (p.43); grooving harness (p.44);
> ice crusher (p.45).

> Catalog of ice wagons, pre-1930 trade magazines of the
> ice business, books or articles of natural ice days,
> photos of people in natural ice scenes (we can make
> copies and return originals), ice toys, window cards
> (25, 50, 75, 100 lbs.), of every ice company, inform-
> ation on any movie (theater, home) in which the days
> of the iceman are shown primarily or in the background.

We buy, trade (duplicates available) or accept donations. W
pay all shipping costs, and can arrange pickup. For thos
who would like to read further on the subject, we are curr
ently preparing a booklet, with many pictures, on the day
of the iceman.

Robert C. Siegel, Jr.
Director